T0203760

PROCEEDINGS OF THE 30TH INTERNATIONAL GEOLOGICAL CONGRESS
VOLUME 5

CONTEMPORARY LITHOSPHERIC MOTION
SEISMIC GEOLOGY

Proceedings of the 30th International Geological Congress

PROCEEDINGS OF THE
30TH INTERNATIONAL GEOLOGICAL CONGRESS

BEIJING, CHINA, 4 - 14 AUGUST 1996

VOLUME 5

CONTEMPORARY LITHOSPHERIC MOTION SEISMIC GEOLOGY

EDITOR:
YE HONG
INSTITUTE OF GEOLOGY, STATE SEISMOLOGICAL BUREAU, BEIJING, CHINA

CRC Press
Taylor & Francis Group
Boca Raton London New York

CRC Press is an imprint of the
Taylor & Francis Group, an **informa** business

First published 1997 by VSP BV Publishing

Published 2019 by CRC Press
Taylor & Francis Group
6000 Broken Sound Parkway NW, Suite 300
Boca Raton, FL 33487-2742

© 1997 by Taylor & Francis Group, LLC
CRC Press is an imprint of Taylor & Francis Group, an Informa business

First issued in paperback 2019

No claim to original U.S. Government works

ISBN 13: 978-0-367-44811-0 (pbk)
ISBN 13: 978-90-6764-269-9 (hbk)

Visit the Taylor & Francis Web site at
http://www.taylorandfrancis.com

and the CRC Press Web site at
http://www.crcpress.com

Proc. 30th Int'l. Geol. Congr., Vol. 5
Ye Hong (Ed)
© VSP 1997

Preface

This volume presents papers selected from The 30th IGC Special Symposia E- Contemporary Lithospheric Motion and Symposia 14 - Seismic Geology. The Special Symposia E consists of 3 sessions: E-1. Spacial measurement of global crustal movements and measurement of contemporary intracontinental movements; E-2. Contemporary crustal stress field, active faults and folds, dynamic information from seismic and volcanic activities; E-3. Contemporary geodynamic models. The Symposia 14 consists of 4 sessions: 14-1. Earthquakes, paleoearthquakes, and active tectonics; 14-2. Earthquakes and tectonophysical environments; 14-3. Prediction of earthquakes; 14-4. Engineering seismology and mitigation of earthquake hazards. Altogether there are 166 papers presented orally or posted at these 7 sessions. During the conference and afterwards, 22 papers were selected from these 7 sessions for this volume mainly based on the recommendation of conveners of each session.

The selection was made to meet the following requirements: 1) the paper presents new problems or new points of view and approaches to previously recognized problems; 2) it contains especially meaningful results. Due to some reasons, it is very regrettable that there may be still some excellent papers which are not included.

The volume is divided into two parts:

Part 1. Geodynamics and Active Tectonics. This part contains 14 papers. It presents the results on contemporary geodynamic model, crustal stress field, active faults, folds and volcanoes.

Part 2. Earthquake Mechanism and Earthquake Prediction. This part contains 8 papers which discuss the tectonophysical environments of earthquake generation and the methodology of earthquake prediction.

As many people realized now that "we live on the same blue planet". Resources and environment are the two most important conditions for human's survival and development. The advance of geoscience provides the knowledge and principles necessary for the rational use of resources and protection of environments. It is hoped that this volume make a reasonable synthesis for the topics discussed at Special Symposia E and Symposia 14 of the 30th IGC which may benefit the advance of geoscience and the human's future.

I would like to extend my heartfelt gratitude to Ding Guoyu, Ma Zongjin, X. Le Pichon, V. Trifonov, J. Mercier, R. Yeats, Ma Jin, T. Shimamoto, Zhang Guomin, Zhang Peizhen and L. Serva for their recommendation and suggestion in the selection of the papers. I also would like to express my sincere thanks to my colleagues, Chen Guoguang, Zhou Qing and Hao Chongtao, for their valuable contribution to the editing of this volume.

Ye Hong (H. Ye)
Coordinating Editor
Deputy Secretary-General, Geological Society of China
Professor, Institute of Geology, State Seismological Bureau, China
Concurrently, Department of Geology, Beijing University

CONTENTS

Proc. 30ᵗʰ Int'l. Geol. Congr., Vol. 5 pp. 1-7
Ye Hong (Ed)
© VSP 1997

On Global Tectonics and its Dynamics

MA ZONGJIN

Institute of Geology, State Seismological Bureau, Beijing, 100029,China

Abstract

The global distribution of continents and oceans demonstrates a double asymmetry of north-south and east-west. The geoid height has a distribution pattern of first-and second-order anomalies. The global active tectonics can be classified into three first-order tectonic systems: the circum-Pacific tectonic system, the mid-ocean ridge tectonic system, and the continental tectonic system of the north hemisphere. They indicate the asymmetry of north-south and east-west as well. The mechanism of the general westward motion of the global plates is presumably associated with the angular velocity differences between crust, mantle and core. These tectonic features of a global scale stem probably from the heterogeneities of the inner structures left by the early evolution of the Earth and from the combined effects of forces with multiple sources.

Key words: Tectonic System, Geoid, Asymmetric Distribution, Dynamic Loading.

INTRODUCTION

One of the basic tasks of present-day geodynamics is to make descriptions and summaries of geometry and kinematics for the tectonics phenomena of global scales in terms of observable or measurable phenomena and processes, to explore the regularities of the coordinated motions of the tectonic systems under the framework of global scales, and to obtain the unified dynamic interpretations. This paper gives a brief summary on the research results of these aspects.

GLOBAL DISTRIBUTION OF CONTINENTS AND OCEANS AND THE GEOID

The double asymmetry of the global distribution of continents and oceans
The basic features of the solid Earth's surface are determined by the continental and the oceanic geomorphic units. A portion of 70.8% of the total Earth's surface area is covered with oceans. Among the three chief oceans (the Pacific, the India, and the Atlantic ocean) the area of each one is larger than the Eurasia continent. As the largest ocean of the Earth the Pacific ocean and its neighboring seas occupy 35.4% of the total surface of the Earth, it is the largest geomorphic unit of the Earth. The tectonism of the oceanic regions plays a dominant controlling role in the formation of the global tectonic patterns.

The present distribution of continents is not even on the Earth's surface. The 65% of all the continents is located on the north hemisphere. North America, South America, Africa, Asia and the India subcontinent have a triangular shape with vertex angles facing south. While their north portions link each other and converge around the north pole. About 81 % of all the land surface is located on a continental hemisphere at the north with its pole near Spain

(0°E, 38°N). On this hemisphere continents occupy 47% and oceans occupy 53% of its total area. The contrary hemisphere is the oceanic hemisphere at the south which includes 11% continents and 89% oceans in respect to its area. Its pole is at New Zealand, suggesting the north-south hemispheric asymmetry of the global continental and oceanic distribution.

Let it be defined that the hemisphere with the longitude 180° as its axis is the 180° hemisphere and the hemisphere with the longitude 0° as its axis is the 0° hemisphere. Then the 180° hemisphere includes the most portions of the Pacific and a few continents and is equivalent to a oceanic hemisphere. While the 0° hemisphere includes most of the Earth's lands and is equivalent to a continental hemisphere. This is the 0°/180° (east-west) hemispheric asymmetry of the global continental and oceanic distribution.

The geoid height
The orbits of artificial satellites have provided the most precise evidence of the large-scale departures of the Earth from spherical symmetry, i. e. the satellite geoid. Figure 1 shows the form of the geoid, representative of the equilibrium ellipsoid from the satellite data combined with gravity measurements on the Earth's surface[4]. The general feature of this map is that the north and south regions at high latitudes have negative anomalies. The lowest points of the geoid are the south pole (-110m) and the north pole (-60m to -70m). While the regions of positive anomalies concentrate along the regions at intermediate and low latitudes. This is the anomaly distribution pattern of the first-order. In addition the regions at intermediate and low latitudes are also characterized by the intercalated distribution of positive and negative anomaly zones or troughs trending NW. For example the New Guinea positive anomaly zone (+100m), the India ocean negative anomaly zone (-60m), the west Africa positive anomaly zone (+50m), and the north Pacific negative anomaly trough (-40m) can be identified in Figure 1. This is the second-order character of great significance. Another remarkable feature is that except Africa the continents of the north hemisphere are regions of the low geoid while the continents of the south hemisphere are almost all of the high geoid. Although the large scale features of the geoid show little correlation with the surface elevation of the crust (continents versus oceans), the geoid is in part an expression of the deep-seated mantle movements responsible for the global tectonic pattern. The mass anomalies responsible for the geoid features are in the upper mantle at depths of a few hundred kilometer[7].

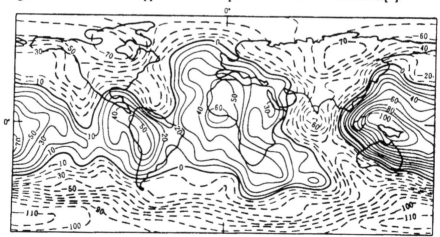

Fig.1 Geoid contours plotted from the coefficients to degree and order (18, 18)[4]. The reference ellipsoid, from which this figure shows the departures, has the equilibrium flattening, 299.76.

THE ASYMMETRY OF THE GLOBAL TECTONIC SYSTEMS

In the light of the global earthquake distribution and its kinematic and dynamic features, the global active tectonics can be classified into three first-order tectonic systems (l) the circum-Pacific tectonic system which is characterized by deep subduction of the oceanic lithosphere under the continental lithosphere; (2) the mid-ocean ridge tectonic system which is marked by the combined tectonics of oceanic lithosphere rifts and transform faults , (3) the continental tectonic system of the north hemisphere which is mainly distributed along the latitudinal ring-shape zone of 20°N-50°N, characterized by the fault network of continental lithosphere and forms four similar regions of seismotectonics (Figure 2).

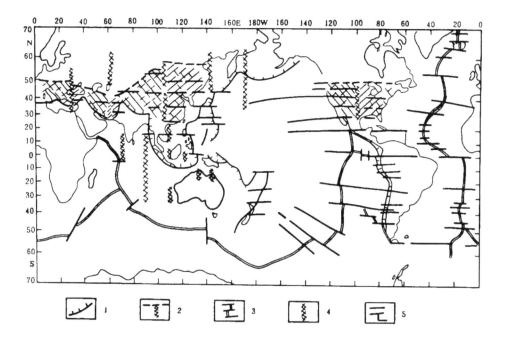

Fig.2 Three tectonic systems of a global scale.
1. circum-pacific tectonic system, 2. continental tectonic system; 3. mid-ocean ridge tectonic system;
4. longitudinal tectonic system; 5. fault.

As mentioned above, for description convenience it is defined that the hemisphere with the axis of the longitude 180° is the 180° hemisphere and the hemisphere with the axis of longitude 0° is the 0° hemisphere. Then almost the whole circum-Pacific tectonic system is located along the outer rim of the 180° hemisphere. The mid-ocean ridge and rift tectonic system, looking like the incomplete framework of a lantern, consists of three longitudinal mid-ocean ridges linked with one latitudinal mid-ocean ridge around the south pole. According to its length the 85% of the ridge axes is distributed on the south hemisphere. They are associated with the south hemisphere which is dominated by oceans, is warmer and has a little dilatation. With a view from the longitudinal direction the three longitudinal mid-ocean ridges and the continental rifts concentrate relatively at the 0° hemisphere, indicating a secondary dilated hemisphere. While the geometrical shape of the circum-Pacific deep subduction zones demonstrates that the 0° hemisphere is overthrusting the 180° hemisphere. The continental tectonic system with its main body of intracontinent orogenic belts within the range 20°-50°N forms a broad, latitudinal ring belt of continental active tectonics. Its regional stress field shows that tectonic deformation of this latitudinal belt is mainly determined by two forces. One is the north-south compression and the other is left-lateral wrench indicated by the spiral-like Earth's surface[5].

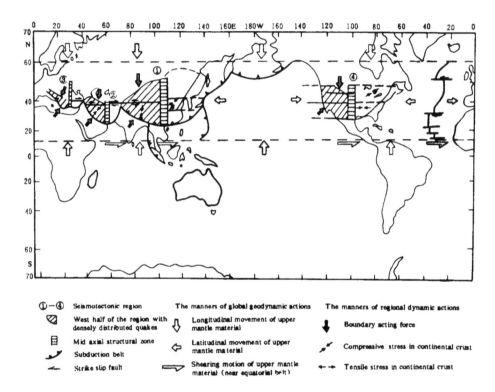

Fig. 3 The continental tectonics of the north hemisphere and its geodynamic background

It is meaningful to compare the disposition of the continental tectonic system and the mid-ocean ridge around the south pole in the spherical coordinate system. The mid-ocean ridge around the south pole occurs on the south hemisphere with a little contraction. Both the two tectonic systems are the latitudinal tectonic belts. This contrast reflects the asymmetry of the thermal regime between the south and the north hemisphere. Another important phenomenon is the asymmetry on the both sides of the roughly north-south (longitudinal) tectonic zones in the three tectonic systems. For example, the west Pacific is characterized by complex marginal tectonics, i. e. integrated trench, island arc and back-arc basin systems. The west-dipping slabs have bigger dip angles (usually greater than 45°). While the east Pacific looks like relatively simple. There are no back-arc basins along its margin. The east-dipping slabs have smaller dip angles (generally less than 45°). But the tiny differential eastward motion of the strip tectonics on the bottom of east Pacific results in regular segmentation of the Cordillera orogenic zone and distribution of earthquakes and volcanoes. The velocities of the spreading ridges, which are indicated by the width of magnetic strips of the oceanic bottom, are often faster on the one side and slower on the other side of one ridge. Within the latitudinal continental tectonic belt of the north hemisphere four seismic regions with similar tectonic deformation patterns can be recognized. All of them can be devided into east and west two half regions by the longitudinal central axis. The west half region has mainly orogenic zones and plateaus with high seismicity and dominant NW trending active tectonics. In contrast the east half-region has mainly planes and hills with relatively low seismicity and dominant NE trending active tectonics (Figure 3). These phenomena demonstrate that the east-west asymmetry with the great longitudinal tectonic zones as their axes is a tectonic feature of the global scale.

DISCUSSIONS ON DYNAMICS

The westward plate motion and the differences of the angular velocities between the crust, mantle and core

The NUVEL-I model of global relative plate motions[1] indicates that the global lithospheric plates are moving generally in a westward direction. There are evident differences of velocities between the plates. When the west plate moves faster than the east plate, there will be extensional fractures such as mid-ocean ridges and continental rifts. Except that both the sides of the east Pacific rise move in the opposite direction, the movements of other plates are decided by the velocity differences of the westward motion. Most of the mid-ocean ridges have north-south trends which are perpendicular to the general westward motion of the plates. When the west plate moves slower than the east plate, then collision or subduction will occur. There may be many kinds of factors which determine the velocity differences of the westward motion, such as (1) undulation of the plate bottom and the differences of the coupling strength between the lithosphere and the underlying mantle, (2) the roughness of the upper surfaces of plates which is associated with the differences of the friction resistance between the continental topography and the atmospheric motion; (3) the differences of the inertia moment produced by the plate masses in the sudden changes of the Earth's axial rotation; (4) the tensional forces by mantle upwelling exerted on the plate westward motion, and (5) the different delaying counterforces on the different plates rising from the solid tide. These possible factors are not related to the mantle convection of large scales.

One of the important inferences of the plate motion model is that the lithosphere is detached from the underlying mantle[2]. Therefore the general westward motion of plates is equivalent to the eastward motion of the mantle. The rate of this relative motion is about 5-10 cm/a. There is also the relative motion between the mantle and the core. The records of geomagnetic basic field for more than 400 years since 1580 indicate its long-term westward drift with the average rate of about 0.2°/yr[3,7] . If the outer core which produces the geomagnetic field is presumably fixed in respect to the inner core, then the westward drift of the geomagnetic field observed on the Earth's surface means that the axial rotation of the crust is faster than the core (since the Earth is rotating from west to east) and slower than the mantle (considering the westward motion of the plates in respect to the underlying mantle). From the above discussions the relative motion between the plates and the underlying mantle seems definite. With another reference system the eastward motion of axial rotation of the mantle is faster than the lithospheric plates, which accounts for the asymmetry on the both sides of the longitudinal tectonic zones mentioned above. The figure of the NUVEL-I plate motion model also demonstrates that the general westward motion of the north hemisphere is faster than that of the south hemisphere. In other words the eastward axial rotation of the lithosphere of the north hemisphere is slower than that of the south hemisphere. Thus a left-lateral wrench belt is formed along the low-latitudinal zone near the equator. This dynamic environment makes directly the India plate, the Arabia plate and the Africa plate push obliquely from southwest towards northeast, resulting in the three intracontinent arc-shaped mountain range systems, i.e. the Alpine, the Zagros, and the Himalaya mountains. The evidence of the westward motion differences between the north and the south hemisphere can be found not only in the low-latitudinal zones but also in the middle-and high-latitudinal zones. The existence of the northwest trending belts in the figure of the geoid height (Figure 1) is a good proof. This wrench of a global scale between the north and the south hemisphere must produce a NE-SW wrench-pressure which is probably associated with the motion of the India-Australia plate with a higher rate in NNE direction and the motion in SSW direction of northwest Canada (Figure 3).

Multiple sources of the forces responsible for the Earth's movement
In the light of the knowledge on the present layering structure of the Earth represnted by the crust, the mantle, and the core as well as on the origin of planets, it is inferred that 4 billion years ago the Earth experienced the following early evolution processes: accretion of cosmic cloud-dust, thermal accumulation of the Earth (the temperature was as high as 1000°C), differentiation and core formation of the iron elements, general melting with increasing temperatures, density layering, and the formation of the primitive crust. It is evident that the common action of gravity and geotherm was dominant during this evolution process. But it must be kept in mind that the early evolution and formation of the Earth was under the status of its axial rotation with a high rate. In the Silurian Period of 440 million B.P. one year was 407 days. It follows that an estimate of the axial rotation of the Earth was equivalent to 700 days per year. Thus the Earth when it was under the state of melting and axial rotation of high rates could have produced the lateral differentiation of some elements such as those belt-shaped structures observed on the surface of the Sun and the Saturn. This prehistory structure of the Earth and its mechanical effects might leave some trace in the distribution of the oldest rocks. At least the old rock strata in some shield regions are dominated by the latitudinal tectonics followed by longitudinal and oblique rock zones. This pattern of the old rock zones may be related to the Earth's structures of pre-geological history.

The north-south asymmetry of the planets has widespread expressions. Even the amount and distribution of the stellars above and below the Milk Way system of dish-shape is also asymmetrical. It is reasonable to suspect that the orientational forces such as the tidal forces outside the domain of the Milk Way system can produce somewhat eccentric effect leading to deviation of the mass center of the star body inside the domain from its geometric center. If this inference is valid then the evolution and movement of the Earth were in progress under the common action of multiple forces including gravity, thermal, axial rotation, tidal and other forces from the beginning. The mass center of the Earth is likely situated on the side of the north hemisphere while its thermal center on the side of the south hemisphere. This disposition determines a series of global asymmetric tectonics.

The present state of the Earth's structure, tectonics and its motion are also determined by the common action of the multiple forces mentioned above. Since the static asymmetric distribution of the Earth's thermal regime and gravity field determines the Earth's shape and the asymmetric distribution of the three tectonic systems as well as the swingwise adjusting motion of hemispheric dilatation and contraction, the giant energy driving the motion of the Earth's tectonics stems presumably from the great plume of thermal pool of the lower mantle and the up welling of the layering structure of the upper mantle coupling with the adjusting action of gravity. It is necessary to consider the orientational nature of the Earth's axial rotation and its inertial force due to changes of the axial rotation velocity during the action process of thermal-gravity forces. It would determine the orientation of a series of tectonics and structures just like the role of the steering weel of a car. Besides the changes of the

Earth's axial rotation can also control the variation of the Earth's flatting and its guiding effect on the motion towards and departing from the poles, in particular to the relative motion between the crust, the mantle and the core due to differences of their axial rotation velocities. Finally it should be mentioned that the astronomical factors of various periods represented by the tidal force, even the strike of giant aerolites, may produce effects of long and short-term adjusting or exciting at any time upon the stress field which is formed under the combined action of gravity, thermal, and axial rotation forces. These astronomical factors can also trigger some abrupt processes such as earthquakes, volcanoes, and magmatic upwelling.

Acknowledgements

This study is supported by the national project: Recent Crustal Movement and its Dynamics and National Foundation of Natural Sciences (49272139).

REFERENCES

1. C. Demets, R. G. Gordon, D. F. Argus and S. Stein. Current plate motions. *Geophys. J. Int.*, 101: 425-478 (1990).
2. C. Doglioni, The global tectonic pattern. *J. Geodynamics*, 1 2: 2 1-38 (1990).
3. Fu Chengyi. 1976. *Ten pieces on the Earth*. Beijing: Science Press (in Chinese).
4. B. M. Gaposchikin, Earth's gravity field to the eighteenth degree and geocentric coordinates for 104 stations from satellite and terrestrial data. *J. Geophys. Res.*, 79: 5377- 5411(1974).
5. Ma Zongjin and Chen Qiang. Global seismotectonic systems and Earth's asymmetry. *Science in China (B)*. 33(1): 121-128 (1990).
6. F. Press, and R. Siever, *Earth*, W. H. Freeman and Company, San Francisco, (1982).
7. F. D. Stacey, *Physics of the Earth*. John Wiely & Sons, New York, (1977).

Proc. 30ᵗʰ Int'l. Geol. Congr., Vol. 5 pp. 9-22
Ye Hong (Ed)
© VSP 1997

Active Faults and Recent Geodynamics of Eurasia

VLADIMIR TRIFONOV, GENNADIY VOSTRIKOV, ROMAN TRIFONOV AND OLGA SOBOLEVA

Geological Institute of Russian Academy of Sciences, 7 Pyzhevsky, Moscow, 109017, Russia

Abstract

The paper describes main results of studying active faults of Eurasia in process of realization of the Project II-2 "World map of major active faults" of the International Lithosphere Program. General regularities of active faulting in the continent: wide belts of faults in the plate boundary areas, predominance of strike-slip motion, manifestations of recent detachment tectonics as well as methods and results of recent geodynamic parameters calculation in the mobile belts are discussed.

Keywords: active faults, Holocene, Late Pleistocene, detachment tectonics, tensor of rates of deformation

INTRODUCTION

The synonyms "active fault" and "living faults" were introduced in the forties by American and European authors, respectively, to designate the ruptures associated with tectonic movements that occur presently or can occur in the nearest future. Since tectonic movements on the faults are nonuniform in time, there arose the problem of a characteristic time which is adequate for specifying the fault activity, direction, mean rate and regime of movements, and associated natural phenomena. In the mobile belts this time interval is Late Pleistocene and Holocene, i.e., approximately the last 100,000 years [1,2]. But it is not enough to estimate parameters of active faulting in the platform regions, where the intensity of fault motion is much lower, and activity pulses related to strong earthquakes are much fewer, than in the mobile belts. Therefore, determination of activity of the platform faults should be based on motion that occurred not only over the last 100,000 years, but also during the Middle Pleistocene, i.e., over the last 700,000 years [3].

In 1989, in view of the significance of active fault studies, the International Lithosphere Program initiated the Project 11-2 "World map of major active faults" with V.G.Trifonov as a chairman [4]. The unique database of manifestations, parameters and seismic effects of active faulting has been created by the Project participants. The largest progress have been achieved in Eurasia. The preliminary map of active faults of the continent has been compiled in scale of 1:5,000,000, and the more detailed maps and descriptions have been compiled for many regions.

This paper represents the results of the geodynamic analysis of active faults of Eurasia. General regularities of active fault position and displacements as well as results of calculation of recent geodynamic parameters by using active fault data are discussed.

STRUCTURAL POSITION AND GENERAL CHARACTERISTICS
OF DISPLACEMENTS OF ACTIVE FAULTS OF EURASIA

An analysis of offsets on active faults of Eurasia showed that the vertical component of displacements is often produced by thrust or reverse motion rather than by normal one. This is true for the faults both in mobile belts and in areas of moderate and weak mobility. Thus, most part of the continent is now under additional lateral compression, which is consistent with estimates of the present-day stress, obtained by different methods[5].

More than half of active faults in the mobile belts of Eurasia have a strike-slip motion component, which is comparable with or greater than the vertical component. It is in strike-slip zones that the highest rates of the intracontinental movements are mod often observed, which may be explained by the fact that the strike-slip movements are less energy-consuming than the motion on thrusts, reverse and even normal faults. We correlated magnitudes of the strongest continental earthquakes of the last decades and a work produced by each of them. The latter was represented by the length of seismic rupture and the length multiplied by the maximum seismic displacement. These characteristics of the strike-slip seismic faults were higher than of the other ones produced by the earthquakes of the same magnitude [6].

The map of active faults of Eurasia with rates of motion not less than 1 mm/year (Fig.1) demonstrates that the longest faults contour recent plate and microplate boundaries. The boundaries are often represented not by single faults, but by belts of active ruptures. Majority of active faults are concentrated in such belts and mobile zones within them. These belts and zones are seen better in the map of all active faults independently on rates of movements on them (Fig.2).

The active (mobile) zones of the orogenic belts (for example, of the Alpine-Asian belt) are often manifested in topography by ridges or systems of ridges, and the stable blocks between them form relative depressions with recent sedimentation. Being narrow in areas of active zone convergence, the stable blocks form intermountain basins. It is just the origin of the basins between Pamirs and Tien Shan in the continuation of the Tarim microplate. Within the mobile zones, active faults bound the smaller uplifted (ridges) and subsided (basins) neotectonic forms. The basins correspond often to the areas composed by the denser rocks (for example, the ophiolites with ultrabasic bodies and their fragments), than adjacent ridges. The land surface of such basins was subsided isostatically before the neotectonic movements. So, many and perhaps all intermountain basins of the recent compressed orogenic belts can not be identified as synclines of the basement in the Argand's [7] terms. They are predetermined by the basement heterogeneity and correspond to remnants of the stable blocks between the mobile zones or to areas of the denser crust within the mobile zones.

The topographic contrast of mountains and intermountain basins increased during the orogeny. The ridges were eroded and became more light-weight, and the basins were filled with classic material and became heavier. This isostatic imbalance was compensated in the depth by rock lateral motion realized in the layers more destructed or plastic and therefore less dense than the adjacent layers [8]. As a result, the basement of the basins enriched by the heavier components independently on the basin origin.

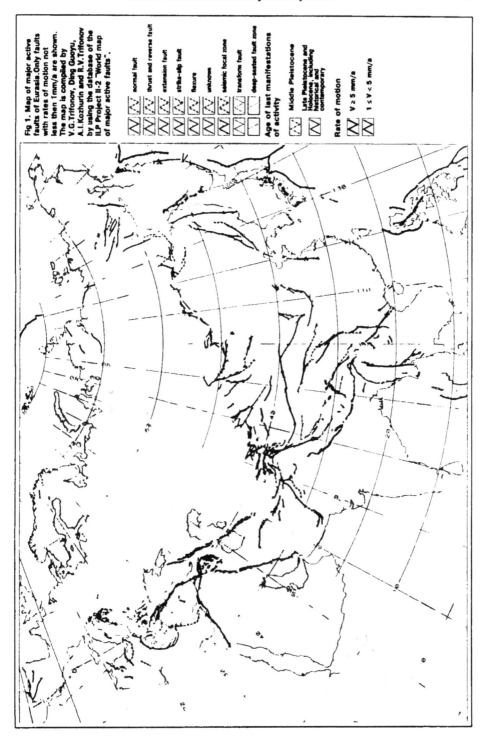

Fig 1. Map of major active faults of Eurasia. Only faults with rates of motion not less than 1mm/a are shown. The map is compiled by V.G.Trifonov, Ding Guoyu, A.I.Kozhurin and R.V.Trifonov by using the database of the ILP Project II-2 "World map of major active faults".

normal fault

thrust and reverse fault

extension fault

strike-slip fault

flexure

unknown

seismic focal zone

transform fault

deep-seated fault zone

Age of last manifestations of activity

Middle Pleistocene

Late Pleistocene and Holocene, including historical and contemporary

Rate of motion

V ≥ 5 mm/a

1 ≤ V < 5 mm/a

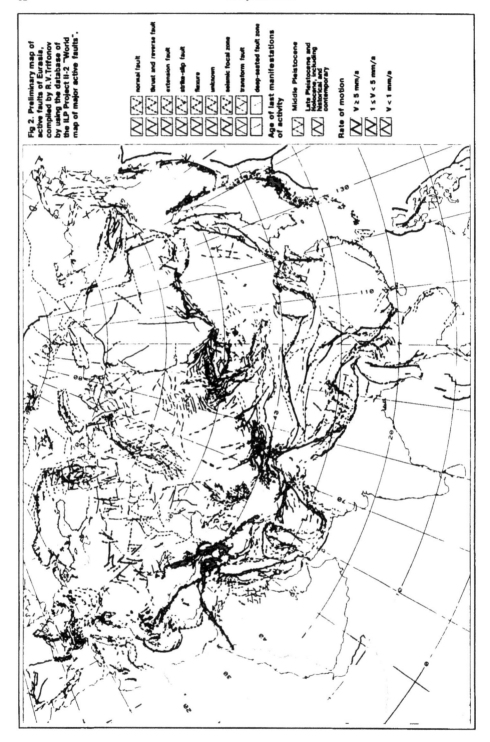

Fig 2. Preliminary map of active faults of Eurasia, compiled by R.V. Trifonov by using the database of the ILP Project II-2 "World map of major active faults".

normal fault

thrust and reverse fault

extension fault

strike-slip fault

fissure

unknown

seismic focal zone

transform fault

deep-seated fault zone

Age of last manifestations of activity

Middle Pleistocene

Late Pleistocene and Holocene, including historical and contemporary

Rate of motion

V ≥ 5 mm/a

1 ≤ V < 5 mm/a

V < 1 mm/a

ACTIVE DETACHMENT TECTONICS

Majority of active faults, identified in the land surface cut only the Upper crust and are not fixed in the lower lithosphere, being deformed by the other way. Evidences of the recent detachment tectonic processes were published in [8,9]. In contrast to the older detachment tectonic features, exposed by the erosion the recent ones can not be observed directly. They are manifested by a concentration of hypocentres of earthquakes in subhorizontal zones and by a difference of structural pattern and style of active faulting in different layers of the lithosphere. In this case, the deep-seated active structural elements can be identified by the related geophysical and geochemical anomalies. The deep-seated elements are manifested sometimes in the land surface by structural features discordant relative to the surficial active structures. The discrepancy of active structures in different depths is caused mostly by the different response of rheologically different rocks to essentially identical loading. However, the orientation of stress and direction of motion are not the same in different layers of the lithosphere in the central Japan, the western

Tien Shan and some other regions. The most complicated active detachment tectonic features have been found in the compressed mobile belts. Their manifestations depend on structural position of the region.

The subduction is characteristic only of the oceanic or suboceanic lithosphere and is commonly associated in the eastern margin of Asia with more or less considerable low-angle underthrusting of fragments or reworked material of the subducted slab beneath the crust of the allochtonous plate [10]. Arcs of the Crete and Lesser Antilles type represent a particular case, because the subduction here is accompanied by the overriding of the allochtonous plate or only its crustal part. In the Aegean region, the latter is caused by the lateral compression of the Anatolian microplate moving westward and by the concurrent extension due to the rise of the anomalous mantle initiated by the tectonic destruction of the regional lithosphere. This mechanism was proposed after the analysis of active faulting [3] and was confirmed by the interpretation of the GPS measurements[11]. Similar overriding is likely to occur in the Pacific island arcs [12], but there it is much dominated by the subduction processes.

The typical collisional interaction in areas of the maximum plate convergence and compression is decoupling and independent deformation of the crust and mantle, sometimes with layering of the crust into several slices (the Punjab-Pamir and the Arabian- Lesser Caucasus syntaxes). Active tectonics of the Upper crust is principally similar in the both syntaxes. Nappe-fold structures develop here. Active strike-slip faults are widespread. Along with strike-slips of transition, which displace one fault side as a whole relative to another (the faults in the western and northeastern boundaries of the Arabian and the Indian plates), the strike-slip faults of rotation and squeezing of rocks away from the area of maximum compression are discriminated [6]. The latter two types of the strike-slip faults produce the lateral shortening of the collision belt and redistribute rocks along it.

A behavior of the Lower crust and the mantle part of the lithosphere (mantle lid) is different in the syntaxes under discussion. In the Pamir-Himalayan region, the mantle lid, with overlying crust detached, being depleted, eclogitizated in some places, relatively cold and therefore heavier than surrounding formations, sinks into the mantle undergoing considerable deformation as well [13]. The mantle focal zone is fixed here up to 270 km to the depth. The Lower crust, being the boundary zone between differently deformed the Upper crust and the mantle lid, is characterized by the most contrast motion and intense deformation.

In and around the Arabian-Lesser Caucasus syntaxis, the mantle lid is abnormally heated and probably enriched by products of the deep mantle differentiation. The heated area is fixed by the young transverse volcanic belt, known in the northern part as the Transcaucasus uplift [14]. The Late Quaternary volcanic belt intersects the heterogeneous structural zones from the Precambrian Arabian plate in the south up to the Great Caucasus in the north. The rift-type basaltic volcanism with relatively shallow (30-40km) magmatic facies took place in the Arabian plate with thickness of the Earth's crust about 35 km. Northward the belt intersects the zones of intense alpine deformation with the variable (30-40km) Earth's crust and possibly the eclogitizated slabs of the Mesozoic oceanic crust of the Tethys beneath it. The Late Quaternary volcanism belongs here to the calc-alkaline series with a predominance of andesites. In the Great Caucasus, where the thickness of the Earth's crust is increased up to 45 km by underthrusting of the southern zones, the acid volcanism is essential [15]. We can suppose by these data that the roof of the deep-seated heated zone is situated in depths of 30-40 km and the zone occupies not only the mantle, but also the Lower crust in the northern part. Presence of such zone made easier the detachment of the crust and its squeezing to the north. Finally it overthrusted completely the Paratethys basins that had existed in the Caucasus up to the Earlier Miocene. The rebuilt fragments of this system of basins are preserved to the west and the east of the Caucasus as the Black Sea and the Southern Caspian. The heated, enriched by products of magmatic differentiation and therefore relatively light-weigt mantle lid has not sunk into the mantle in the Caucasus region. But immediately to the east of it, in the Apsheron threshold, the dipped to the north mantle focal zone is fixed up to 100 km to the depth.

A "bulldozing" is a main mechanism responsible for the distribution of deformation and motion to the north of the regions of maximum recent collision. It is characteristic over vast areas of the Central and Eastern Asia. The mechanism is related here to the northward drift of the Indian plate, giving rise to the deformation and motion of adjacent microplates and crustal blocks, which in their turn cause the tectonic zones adjacent to them to move and so on. The deformation and motion are concentrated in the boundary zones between the microplates and blocks and essentially depend on their configuration. The intensity of neotectonic folding and detachment diminish to the north and northeast of the regions of maximum collision and are replaced by purely fault-type structures dominated by strike-slip movements. Extension structures (the Baikal and Shanxi rift systems) develop on the curved segments of large shear belts.

In the region of Alpine Europe and Mediterranean, the processes of recent crustal detachment and "bulldozing" take place, but are less pronounced, being of local character and associating with the lithosphere extension features. The latter are represented both by rift zones and by isometric basins such as the Pannonian and Aegean basins. The wide occurrence of such structures may be related to the mantle diapirism initiated by the interaction of plates and blocks in the environment of much thinner crust and more heated lithosphere as compared with the Central Asia.

RECENT UPPER CRUST GEODYNAMICS BY USING ACTIVE FAULT DATA

The Central Asian collision region between 26°and 56°N and 64°and 104°E is under discussion (Fig.2). It occupies the Tien Shan, Altai, Sayan, Pamir, Hindu Kush, Kunlun, western Himalaya mountains and adjacent territories of Afganistan, Pakistan, western

Mongolia and western China, including Tibet.

We have used parameters of active faults for calculating the recent geodynamic field and have not taken into account the deformation produced by active folding because of difficulties of its measurement and relatively small contribution to the total deformation. The database of active fault parameters have been formed by using the data on active faults of the region, collected in a process of realization of the Project II-2 "World map of major active faults" of the International Lithosphere Program. The papers of Ding Guoyu, N.V.Lukina, P.Molnar, T.Nakata, A.A.Nikonov, V.P.Solonenko, P.Tapponnier, V.G.Trifonov, K.E.Abdrakhmatov, Deng Qidong, V.S.Burtman, S.D.Khilko, K.G.Levi, V.I.Makarov, S.I.Sherman, A.Sinha, A.V.Timush, and R.S.Yeats have been used at first for the database.

The database includes the following parameters of every fault: (1) number and name (if the latter exists); (2) source of information; (3) location, represented by necessary number of points with defined geographic coordinates; (4) direction of dip of the fault plane: NO (for NE, N, NW, and W) or ZU (for SW, S, SE, and E); (5) angles of the dip, degrees: min-mp-max; (6) sense of lateral component of motion: D (dextral) or S (sinistral); (7) presence of extension component: E; (8) sense of vertical component of motion: R (thrust or reverse) or N (normal); (9) average rate of the lateral motion, mm/year: min-mp-max; (10) average rate of the extension, mm/year: min-mp-max; (11) average rate of the vertical motion, mm/year: min-mp-max.

"Min" means the minimum value of the parameter, "mp" means the most probable value and "max" means the maximum one. The average rate has been estimated for Late Pleistocene and Holocene. If the dip or the average rate had not been defined by field data, we have estimated them in wide interval of possible values by using general neotectonic and seismological dada on the fault area. The average rate can be a result of creep or pulses of displacements during rare strong earthquakes. We have estimated it by using geological, geomorphological and more rarely seismological or geodetic data. If any of the parameters (4-11) varies along the fault, it has been differentiated to monotonous segments.

The "hydrodynamic" model of the Upper crust is proposed, such as: discrete displacements on active faults are represented formally as an elements of unified process of a viscosity fluid flow in a large space-time volume. One of the macroscopic parameters of this process of discrete-continuous flow is the deformation rate tensor. It is calculating as an effective average parameter for the large space-time volumes. This parameter being multiply by viscosity factor is the stress tensor.

Time condition is completely fulfilled, because the Late Pleistocene-Holocene (100, 000 years) has been taken. We take into consideration the layer of the Upper crust only (15km). For fulfillment of space condition, a lateral dimensions of the volume (linear dimensions of the elementary window) must be longer, than the largest fault. On the other hand, a dimension of the window can not be too big, because we consider it as an elementary volume of the flow. Dimensions of the windows, been choosen for a calculation of the average deformation will be specified below.

To begin with, the monotonous segments of faults have been in turn subdivided onto the elementary cells as long as of 10 to 20 km with constant strike and dip. To avoid an attenuation of the displacement amplitudes, the faults' termination has been cut off. The termination do not exceed 5% of the total fault lengths. Width of the cell (the fault penetration depth) is correlated with the fault length. According to [16]:

$$\lg L_3 \,(\text{km}) = 0.75 \cdot \lg L - 0.07, \text{ if } L_3 \le 50 \text{ km}, \tag{1}$$

where L is a total fault length and L_3 is a width of the cell. L_3 can not exceed 15 km. We introduce the value:

$$M = S \cdot L_1 \cdot L_3, \tag{2}$$

where S is a displacement vector amplitude along the cell and L_1 is a length of the cell. We call M as a "geometrical moment". By the sense this value, being multiply by coefficient of dry friction on the fault is the moment of the force acting to the cell analogous to the seismic moment of an earthquake source [17]. For any elementary window one may to introduce the orthogonal coordinate system X,Y,Z (east, north and zenith respectively). Then the moment geometrical tensor define by expression:

$$M_{lm} = M(l_s \cdot m_n + l_n \cdot m_s), \tag{3}$$

where l, m = x,y,z; and l_n, l_s, and m_n, m_s are cosines of the direction of the local coordinate system (n is a normal to the cell plane and s is a direction of the vector of displacement along the cell). Within each window all similar components of the tensor (3) are summarized and normalized to a units of the window volume and time:

$$\dot{\varepsilon}_{lm} = \frac{1}{2} \frac{\sum\limits_{n} M_{lm}^{(n)}}{\Delta V \cdot \Delta T}, \tag{4}$$

where n is an amount of the cells inside any window; (ΔV is a square of the window multiplied by the thickness of the active layer (15 km); (ΔT is the Late Pleistocene and Holocene time interval. After B.V.Kostrov [17], $\dot{\varepsilon}_{lm}$ is an average tensor of deformation rate at the expense of motion along active faults.

Using the well-known technique of the rock mechanics [18], the principal rates of deformation (M_1, M_2 and M_3) are calculated and prescribed to the window center.

The region under consideration is divided along the geographic parallels and meridians onto the elementary windows by two ways: at first, of size of $1° \times 1.25°$ without overlapping and at second, of size $3° \times 3.75°$ with a step of $1°$ and $1.25°$ correspondingly. The first differentiation shows more detailed picture and is used for calculation of the direction of principal deformation only. The second differentiation shows the smoothed picture and is used for calculation both direction and magnitudes of principal deformation. As a result, a field of deformation rate tensor has been shown by the directions of principal shortening and lengthening (Fig. 3, 4) and by isolines of their magnitudes (Fig. 5, 6).

Axes of principal shortening (Fig. 3) are subhorizontal all over the region and are approximately north-trending over the most part of it. It is more clear on the smoothed picture, but the residuals from this general direction are better seen on the detailed map. They are manifested on both sides of the Pendjab syntaxis: in the eastern part of the Tadjik basin, in Baluchistan and partly in the northeastern Afganistan to the east of the syntaxis and more noticeable in the active zones to the east of the syntaxis, such as the strike-slip zones of Altin Tagh, the northern Tibet, Gobi Altai, Khangay (the northern Mongolia) and the northeastern

Sayans. In the eastern framing of the Tibet and the Qaidam the axes of principal shortening are almost east-trending.

The axes of principal lengthening (Fig. 4) are also subhorizontal in this "anomalous" zones that shows strike-slip motion along them. The more complicated picture is observed in the eastern part of the Tadjik basin and in Balushistan, where the axes of principal lengthening are subvertical. It corresponds to thrust motion.

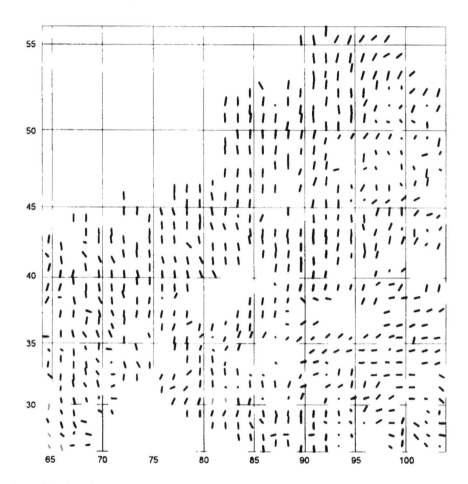

Fig. 3 The orientations of principal shortening in the region. Length of lines is proportional to their angle with a vertical.

Both the shortening and the lengthening axes are subhorizontal in majority of the regions with approximately north-trending direction of the principal shortening. It manifests a predominance of strike-slip motion. The latter appears to be the most prevailing type of motion over all the region under consideration. At the same time, the direction of principal lengthening becomes almost vertical in some areas. They are the most part of Himalayas, the western and southeastern Tien Shan, the southeastern Altai and Sayans. Evidently, the prevailing type of motion is thrusting in these areas. It is interesting that subvertical direction of principal lengthening is often observed in some blocks, weakly ruptured by active faults.

They are the northern part of Tibet, the eastern Qaidam, the central and the northern parts of the Dzungarian basin, the Inner (Chinese) Mongolia. The most part of these areas, except Tibet, is represented by the large-scale and weakly deformed intermountain basins. The foot of the Earth's crust and the foot of the Upper crust are uplifted within the basins relative to adjacent mountain systems. The combination of the vertical lengthening and relative subsidence of the land surface in the basins is unusual and demands additional studies.

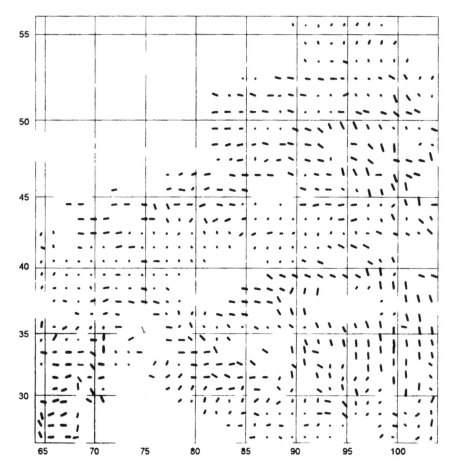

Fig. 4 The orientation of principal lengthening in the region. Length of lines is proportional to their angle with a vertical.

Space distribution of magnitudes of rates of principal shortening M_3 (Fig. 5) and lengthening M_1(Fig. 6) shows their little difference over the large part of the region. It means the deformation has appeared to be double-axes and corresponds to a pure shear deformation in the rock mechanics. However, there are some areas with the larger magnitudes of the principal rates of shortening M_3 and lengthening M_1, and as a rule $M_3>M_1$. The largest magnitudes of M_2, i.e. the most difference between M_3 and M_1 take place in the northern termination of the Kobdo fault zone in the Mongolian Altai, the Talas-Fergana fault zone and adjacent parts of the southern and central Tien Shan, the northern Pamirs (Peter the First Range), the zone of junction of the Chaman and the Herat faults to the south of town of

Faizabad in Afganistan, and the epicentral areas of the Assam and Bikhar earthquakes in Himalayas. The lesser, but also significant magnitudes of M_2 exist in the northeastern flank of Tibet.

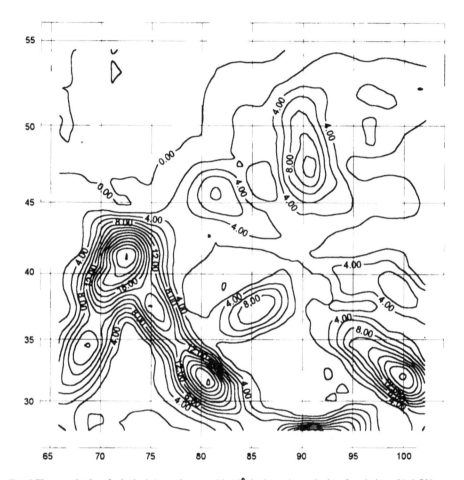

Fig. 5 The magnitudes of principal shortening rate ($M_3 \cdot 10^9$) in the region, calculate for windows $3° \times 3.75°$.

Come to analyze the magnitudes of deformation rates, one can notice that the largest rates are observed in the recent northern boundary of the Indian plate: the Himalayas, the Pamir-Karakorum fault, the northern flank of Pamirs and the adjacent part of the southern Tien Shan (including the central segments of the Talas-Fergana fault), the Darvaz and the Chaman faults. The lesser, but large enough deformation rates are fixed in the eastern boundary of Tibet (the northern Yunnan), the central part of the Altin Tagh and the Hayuan strike-slip zones (i.e. in the northwestern and northeastern flanks of the Qaidam), the northwestern termination of the Dzungar fault and the Mongolian Altai. All these regions are the recent boundary zones between the microplates within the orogenic belt.

CONCLUSION

It has been shown [19] that recent tectonic phenomena and formations are a result of

interaction of open tectonic systems. Such system is a set of natural processes linked in a
particular volume of the geological environment and leading directly or indirectly to
movements in the lithosphere and the development of structural forms. At the same time, a
tectonic system is a structural-stress system developed at some level of organization in the
medium as a result of deviation from the equilibrium state for any of the parameters
characterizing it as a thermodynamic system. The system rank is indicated by the size of the
region in which the links between the elements are closed. In that sense, one can speak of
systems of a global scale and of local ones of various ranks.

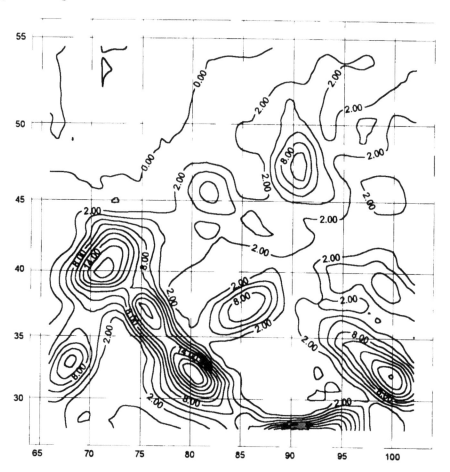

Fig. 6 The magnitudes of principal lengthening rate ($M_1 \cdot 10^{-9}$) in the region, calculated for windows 3°×3.75°.

The global system is the modified version of the plate tectonics that takes into account the
detachment of the lithosphere layers, relatively independent deformation of the separated
layers, and geodynamic and structural results of these processes [20]. The local systems vary
by sizes and structural manifestations. They are initiated by external factors that are the
processes in the global system or in the local one of the larger size. For instance, an uplift of
the abnormal mantle produces deformation of the Upper crust. On the other hand, a
destruction of the lithosphere because of tectonic interaction of the crustal blocks can
decrease a density and produce a heat-up of the mantle with the adequate volcanic and

structural results.

We have discussed the influence of a compensation of isostatic imbalance because of erosion and sedimentation to the development of mountain uplifts and intermountain basins. More complicated interaction of the tectonic systems is manifested by the Holocene faulting in Fennoscandia. The transverse (NW-SE) compression, as a response of the thick continental lithosphere to the Atlantic spreading, interacts with the nonuniform isostatic uplift after a release of the glaciation loading. The linkage of earthquakes and active faults is evident, but we found also a correlation between strong earthquakes and ophiolite zones with ultrabasites in the Arabian-Caucasus region [21]. Perhaps, the transformation of ultrabasite to serpentinite increases the rock volume and produces additional stress participating in the earthquake preparation.

Teilhard de Chardin [22] designated the present time by the momentary cross-section of the unlimited time fibers. The geological structure predetermines many peculiarities of the active tectonics of the region. Similar, but differently developed structures are distinguished by active tectonic characteristics. For example, the Adriatic syntaxis, similar in the Oligocene with recent the Arabian-lesser Caucasus and the Punjab-Pamir syntaxes, was rebuilt later. Its active faulting displays a combination of new structural element and fragments of the syntaxis continuing to develop.

Thus, active faulting is a result of complicated interaction of different tectonic systems being now in different stages of their evolution.

Acknowledgments

The studies were supported by the International Lithosphere Program, the Project 11-2, and the International Science Foundation, the Project MPJOOO.

REPERENCES

1. V.G.Trifonov. *Late Quaternary tectonics.* Nauka Press, Moscow (in Russian) (1983).
2. V.G.Trifonov. Development of active faults, *Geotectonics* 19, 95-103 (1985).
3. V.G.Trifonov. World map of active faults (Preliminary results of studies), *Quaternary Intern.* 25, 3-12 (1995).
4. V.G.Trifonov and M.N.Machette. The World Map of Major Active Fault Project, *Annali di Geofisica* 36, 225-236 (1993).
5. P.N.Kropotkin. Tectonic stress in the Earth's crust, *Geotectonics* 30 83-94 (1996).
6. V.G.Trifonov. General features and peculiarities of the recent geodynamics of the continents.In: *Geodynamics and evolution of the tectonosphere.* R.G.Garetsky (Ed.). pp.144-160. Nauka Press, Moscow (in Russian) (1991).
7. E.Argand. La tectonique de l'Asie. In: 13e Congr. Geol. Int. pp.171-372. *Vaillant-Carmanne*, Liege (1924).
8. V.G.Trifonov, V.I.Makarov, and G.A.Vostrikov. Structural and dynamic layering of the lithosphere in neotectonic mobile belts. In: 27th IGC, Reports 3, *Quaternary Geology and Geomorphology.* pp.105-117. Nauka Press, Moscow (1984).
9. Yu.M. Puslicharovsky and V.G.Trifonov (Eds.) *Tectonic layering of the lithosphere and regional geological studies.* Nauka Press, Moscow (in Russian) (1990).
10. A.I.Kozhurin and G.A.Vostrikov. Kurile-Kamchaka island arc system. In: *Neotectonics and the recent geodynamics of mobile belts.* P.N.Kropotkin (Ed.). pp.67-135. Nauka Press, Moscow (in Russian) (1988).
11. H.Drewes and E.Geiss. Modellirung geodynamischer Deformation in Mittelmeerraum, Satellitengeodasie. VCH Verlagsgesellschaft mbH, D-6940, *Weihelm.* 5.335-349 (1990).
12. I.V.Melekestsev. *Volcanism and relief formation.* Nauka Press, Moscow (in Russian) (1980).
13. A.V.Nikolaev, I.A.Sanina, V.G.Trifonov, and G.A.Vostrikov. Structure and evolution of the Pamir-Hindu Kush region lithosphere, *Phys. Earth and Planet. Inter.* 41, 199-203 (1985).
14. E.E.Milanovsky. *Neotectonics of Caucasus.* Nedra Press, Moscow (in Russian) (1968).
15. E.E.Milanovsky and N.V.Koronovsky. *Orogenic volcanism and tectonics in the Alpine belt of Eurasia.*

Nedra Press, Moscow (in Russian) (1971).

16. A.V.Sidorenko (Ed.) *The map of faults in the USSR territory and the adjacent areas*. Nedra Press, Moscow (1978).

17. B.V.Kostrov. *The mechanism of the tectonic earthquake source*. Nauka Press, Moscow (in Russian) (1975).

18. A.Jeager and G.Cook. Fundamentals of rock mechanics. *Sci. Paperbacks*, N.Y. (1969).

19. V.S.Ponomarev and V.G.Trifonov. Factors of tectogenesis. In: *Actual problems of oceanic and continental tectonics. pp.81-94*. Nauka Press, Moscow (in Russian) (1987).

20. V.G.Trifonov. Neotectonics and current tectonic concepts, *Geotectonics* 21, 18-29 (1987).

21. T.P.Ivanova and V.G.Trifonov. New aspects of correlation of tectonics and seismicity, *Doklady Akad. Nauk* 331, 587-589 (in Russian) (1993).

22. P.Teihard de Chardin. *Le Phenomane humain*. Paris (1958).

Proc. 30ᵗʰ Int'l. Geol. Congr., Vol. 5 pp. 23-33
Ye Hong (Ed)
© VSP 1997

Some Remarks on Seismotectonics and Geodynamics of the Chinese Continent

YE HONG, CHEN GUOGUANG, ZHOU QING AND HAO CHONGTAO
Institute of Geology, State Seismological Bureau, Beijing, 100029, China

Abstract

The contemporary tectonic movement in China is governed by both Tethys-Himalayan and Western Pacific ongoing geodynamic processes. In the west, the tectonic driving force is primarily from the collision between the Eurasia Plate and India Plate. This process also affects much of the East China. Whereas, the southeastern coastal region and the northeastern part of China are mainly affected by the motion of the Philippine Sea Plate and the Pacific Plate, respectively. Under the present-day geodynamic regime, the Chinese Continent is divided into 6 intraplate tectonic blocks: Qinghai-Tibet Block, Gansu-Xinjiang Block, Northeast China Block, North China Block, South China Block and Southeast China Block. Relative movements have been found along boundaries between the intraplate blocks. The tectonic movement between intraplate blocks is usually distributed over a wide area. The mechanism and style of these movements are considerably different from those along interplate boundaries. The crustal deformation inside these intraplate blocks are mainly four types: thrusting and crustal shortening; conjugate strike-slip faulting; localized pull-apart extension and block rotation. The prehistoric earthquake study shows that the recurrence interval of major earthquakes along major active faults in Chinese Continent is commonly much longer than one thousand years. Thus, the strain rate inferred from the earthquake data in Chinese Continent is much lower than that in the interplate regions.

Keywords: seismotectonics, geodynamics, Chinese Continent

INTRODUCTION

Because the two most important tectonic mega-zones: the Circum-Pacific Zone & Tethys-Himalayan Zone conjoin and interact with each other in China, the Chinese Continent provides many spectacular examples of continental deformation: the collision between India and Eurasia Plate; the uplifting, shortening, thickening and extrusion of Qinghai-Tibet High Plateau; the expulsion of South China; the rejuvenation of the Tien Shan and Altar Mountain; The opening and propagating of the North China Cenozoic rifting; The influence of the subduction and collision in Western Pacific margin to the Southeast China and Northeast China (Fig. 1). All these hot topics have attracted the world-wide attention.

Earthquakes are the results of lithosphere deformation under the present-day geodynamic regime. Earthquake data are considered to be one of the basis for the understanding of the contemporary lithosphere deformation. The purpose of this paper is to review and discuss the principal features of the present-day tectonic process in Chinese Continent based on the new results from the study of seismotectonics.

The data we used in this paper include: the earthquake catalog since 780 B.C. to present with magnitude above 4.7 (Fig. 2); the newly revised earthquake focal mechanism solution of 325 earthquakes in China[5], among them 46 focal mechanism solutions are presented in Fig. 5; the surface rupture and ground deformation of 50 major earthquakes and the paleoseismic data from 78 trench studies(Fig.4)[6]. These data were mainly collected and analyzed for authors' previous national seismic zoning work [5, 6, 18, 19].

The mechanism and space-time distribution of earthquakes in China and the relationship between seismicity and tectonics are reviewed first. The intraplate dynamic and kinematic model of active tectonism have been further discussed briefly based on the seimotectonic characters of China.

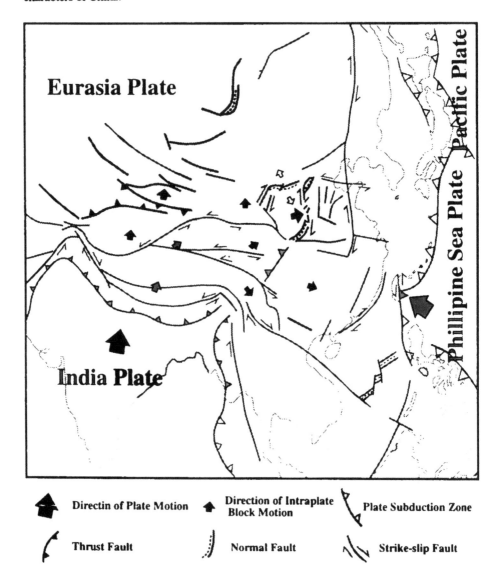

Fig. 1   Present-day Tectonics in China and its Adjacent Regions

SEISMOTECTONIC PROVINCE AND INTRAPLATE BLOCK

Based on the inhomogeneous distribution of the seismicity and the spacial variation of seismotectonic setting (Fig. 2), China can be divided into the following seismotectonic provinces:
1) Gansu-Xinjiang S.P.
2) Qinghai-Tibet S.P.
3) Himalayas S.P.
4) Northeast China S.P.
5) North China S.P.
6) South China S.P.
7) Southeast China S.P.
8) West and Central Taiwan S.P.
9) East Taiwan S.P.
10) South China Sea S.P.
Each of the above-mentioned seismotectonic provinces has a unique contemporary tectonic setting and crustal deformation pattern.

Two of them, the Himalayas Seiemotectonic Province and East Taiwan Seismotectonic Province coincide with the Himalayan Collision Zone and East Taiwan Collision Zone, respectively. The West and Central Taiwan Seismotectonic Province and South China Sea Seismotectonic Province behave as the transition zone between interplate and intraplate region. The other six seismotectonic province, which located in Chinese Continent, are mainly of intraplate nature.

From the comparision between the location of intraplate seismotectonic province and the distribution of Precambrian crystaline basement[14], it appears that the above-mentioned six intraplate seismotectonic province in Mainland China either simply coincides with a Precambrian craton or takes one or two rigid microplatform or massif as their cores. These microplateform or massif are located in the central part of intraplat seismotectonic province and usually surrounded by relatively ductile Paleozoic-Mesozoic fold belts. For example, the North China Intraplate Seismotectonic Province coincides roughly with the well-known Sino-Korea Plateform. The Northeast China Intraplate Seismotectonic Province is composed by Songnen Massif and the surrounding Paleozoic fold belts. The South China Intraplate Seismotectonic Province is made up by the western part of Yangtze Platform and surrounding Paleozoic fold belts. The Southeast China Intraplate Seismotectonic Province coincide roughly with Cathaysian Massif. The Gansu-Xinjiang Intraplate Seismotectonic Province is composed by Tarim Plateform, Junggar Massif and intervening Paleozoic fold belts. The Qinghai-Tibet Intraplate Seismotectonic Province is made up by a series of massifs such as Qaidam Massif, Qiangtang Massif, Gangdise Massif, Songpan-Bikon Massif and Several imbeded Paleozoic-Mesozoic fold belts.

In general, the boundary of seismotectonic provinces takes advantage of pre-existing fault zones or suture zones. But it does not necessarily restrict to the boundary of the former tectonic units. Relative motion along the boundaries of these blocks can be identified by the clustering of seismicity (Fig. 2, Fig. 3, Fig. 4). This pattern suggests that the block faulting plays an important role in the present-day intraplate tectonism in China. The six intraplate seismotectonic provinces in China act as six major intraplate tectonic blocks. The present-day tectonism in Mainland China can be described by the relative motion of these major intraplate blocks and the internal crustal deformation within these major blocks.

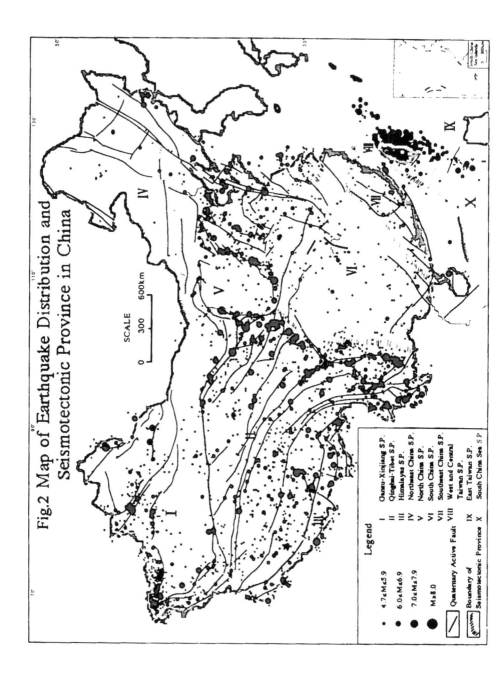

Fig.2 Map of Earthquake Distribution and
Seismotectonic Province in China

Legend

I	Gansu-Xinjiang S.P.
II	Qinghai-Tibet S.P.
III	Himalaya S.P.
IV	Northeast China S.P.
V	North China S.P.
VI	South China S.P.
VII	Southeast China S.P.
VIII	West and Central
	Taiwan S.P.
IX	East Taiwan S.P.
X	South China Sea S.P.

$4.7 \leq M \leq 5.9$
$6.0 \leq M \leq 6.9$
$7.0 \leq M \leq 7.9$
$M \geq 8.0$

Quaternary Active Fault

Boundary of
Seismotectonic Province

SCALE
0   300   600km

BLOCK MOTION AND DRIVING FORCE

As indicated by the paleoseismic data, in some area the slip-rate between two intraplate blocks can be up to 10-20mm per year [2,13,15,22]. For example, the well-known northeast-east trending left-lateral strike-slip Altyn Tagh Fault acts as the boundary between Qinghai-Tibet Intraplate Block and Gansu-Xinjiang Intraplate Block This is a relatively narrow strike-slip zone featured with clear surface ruptures caused by Holocene paleoseismic events. The existence of this type of block boundary indicates that in some part of continental interior major fault zones play an important role in the crustal deformation, somewhat similar to that in the interplate regions.

But commonly, in contrast to the interplate boundaries, the seismicity along most of these intraplate boundaries are distributed over a wide area. For example , along the eastern margin of the Qinghai-Tibet Block, the seismicity is distributed over a wide area and there is clear discordance between the surface fault zones and seismicity clustering (Fig.2). This suggests that the upper crust and the lower crust there may detach from each other and behave in different way. The crustal deformation is complicated there and does not concentrate on one or two major faults. In some region, the seismicity along the block boundary is quite week, as we see in Fig. 2 on the northern margin of the North China Intraplate Block. This type of fuzzy block boundary reflects the coherent nature of the continental deformation..

It is evidenced by the focal mechanism solutions (Fig. 3) and earthquake surface faulting data (Fig. 4) that the contemporary tectonic movement and deformation of these intraplate blocks are mainly driven by the forces from west and east plate boundaries.

In the western part of China, the tectonic driving force is primarily from the collision zone between Eurasia and India Plate (Fig. 1). The Qinghai-Tibet Intraplate Block strongly pushed by India Plate is undergoing crustal shortening, thickening and uplifting. At the same time it is moving northward relative to the eastern part of China. This causes north-south trending right-lateral shearing between West China and East China resulting into the most notable N-S trending clustering of seismicity (Fig. 2). In China we call it N-S trending Seismic Belt.

Meanwhile, the Tethys-Himalayan Mega-tectonic Zone terminates in the middle of China .Due to this termination the north-south compression in western part of China is asymmetrical . This results into the eastward extrusion of the Qinghai-Tibet Intraplate Block and consequently the southeastward expulsion of South China Intraplate Block[20]. The mechanical behavior which accommodates the eastward extrusion of Qinghai-Tibet Intraplate Block and the right lateral shearing between West China and East China is definitely very complicated. Since the seismicity is distributed over a wide area and inconsistent with the surface faulting, it seems reasonable to assume that the upper crust deformation and the deep lower crust deformation are detached and the block rotation would happen in the upper crust to accommodate the right-lateral shearing in the lower crust.

The driving force from the west also affects much of the East China. The uniform NEE-SWW trending compression stress in North China and the E-W to SEE-NWW trending compression stress in South China are mainly contributed by this process (Fig3). But our earthquake data suggest that the southeastern coastal region and northeastern part of China are mainly affected by the motion of Pacific Plate and Philippine Sea Plate, respectively (Fig.

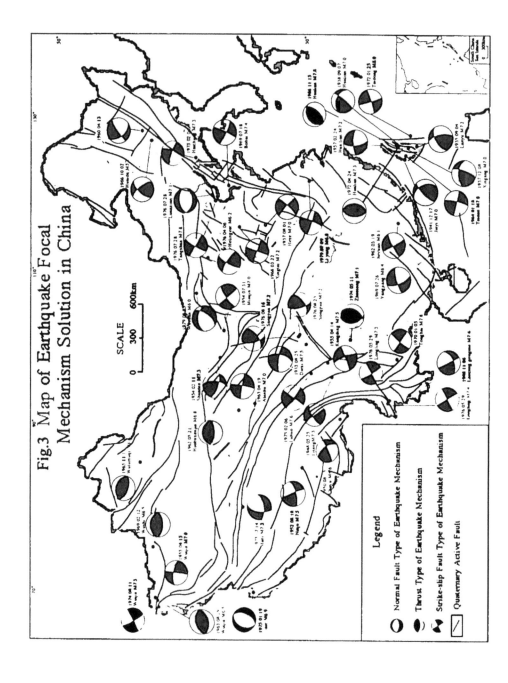

Fig.3 Map of Earthquake Focal
Mechanism Solution in China

2-4). The present-day tectonic setting in northern China is rather complicated. West of the Taihang Mountain, the contemporary tectonic movement is caused by the uplifting and east-northeastward movement of Qinghai-Tibet Block. Whereas in the region east of the Tanlu Fault Zone, the Western Pacific Plate motion seems to play an important role. The region between Tanlu Fault Zone and Taihang Mountain behaves probably as a transition zone interacted by both sides.

INTRABLOCK DEFORMATION

Earthquake data show that there are 4 types of crustal deformation which play an important role in the internal deformation inside these blocks.

1. thrusting and crustal shortening
In the western part of China, the thrusting and reverse faulting associated with the crustal shortening and thickening plays an important role in accommodating the convergence between plates and intraplate blocks. Many major earthquake events in the west are thrusting or oblique thrusting, like 1965 Wulumuqi Earthquake, 1969 Wushu Earthquake,1985 Wuqia Earthquake. (Fig.3)It is inferred that this type of deformation accounts for the rejuvenation of the Tien Shan and Altar Mountain[1].

2. Strike-slip faulting
In the eastern part of China, the strike-slip faulting appears to dominate the region strain field. earthquake takes advantage of pre-existing weak zones which are favorably oriented for the strike-slip faulting to accrue. For example, in North China Intraplate Block because the compression stress is treding NEE, therefore the pre-existing NNE trending faults and NWW trending fault are most earthquake-prone, making a regional conjugate strike-slip pattern. Among them the rifting zones with younger age are most favorable seismogenic structure where the crust was considerably necked and weakened, For example, the 1966 Xintai earthquake occurred along the NEE trending Shulu tertiary rifting and necked zone[21]. Blind earthquake sources with focal depth ranging 10-15km are commonly found in Chinese continent. In the eastern part of China, in most cases the earthquake-causitive fault does not reach the surface structural layer. This implies the decoupling between structural layers with different ages and reflects the complexity of continental crustal deformation.

3. Pull-apart extension.
Pull-apart extension caused by major strike-slip faulting are commonly found in the East China, especially in its western part. The best example is in the western part of North China Block where the right-lateral N-S trending led to the opening of Neogene-Quaternary grabens[17]. The focal mechanism solution of 1979 Wuyuan Earthquake indicates a typical normal faulting (Fig. 3). The paleoseismic study also clearly show normal faulting type of deformation.[15]

4. Block rotation
Block rotation accompanied by strike-slip faulting also plays an important role in continental crustal deformation. A notable example is in the east part of Qinghai-Tibet Intraplate Block where a series of NW-SE trending left-lateral strike-slip faults parallel to each other cutting the whole area into several small blocks[7]. In that area the upper crust clock-wise rotation accommodates the lower crust right-lateral shear movement and eastward extrusion resulting into the regional rapid uplifting and crustal thickening.

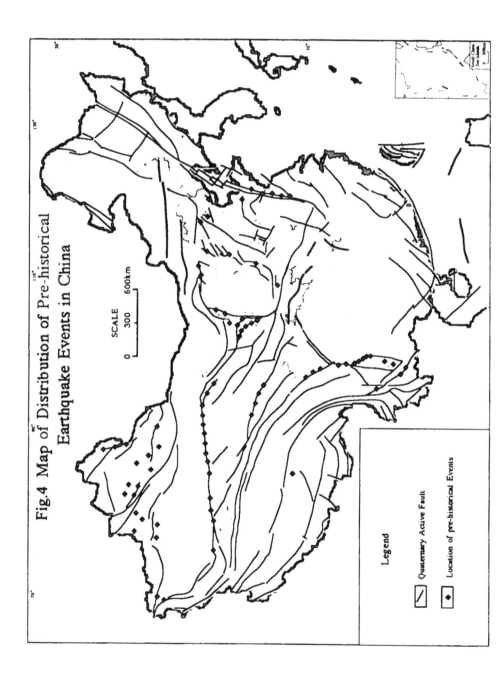

Fig.4 Map of Distribution of Pre-historical Earthquake Events in China

SCALE
0 300 600km

Legend

Quaternary Active Fault

Location of pre-historical Events

EARTHQUAKE RECURRENCE AND SLIP RATE

The results from the study on the prehistoric earthquakes indicate that the recurrence interval of major earthquakes along major active faults in Chinese Continent is much longer than that in the interplate regions [4]. The average recurrence interval in Qinghai-Tibet Intraplate Block and its marginal regions is estimated to be about 1000-2000 years, in Gansu-Xinjiang Intraplate Block 2000-3500 years and in North China Intraplate Block about 2000-4000 years (Tab.1). The slip rate inferred from the above-mentioned estimation places critical constraint on the kinematic model of active tectonism in Chinese Continent. It would be of interest to note that the slip rate inferred from the earthquake data is slower than that anticipated by some previously-proposed kinematic model[12]. This requires further study.

Tab. 1 The recurrence interval of pre-historical earthquake events in mainland China

Fault Name	Number of Events	Magnitude	Average Interval(year)	Maximum Surface displacement (m)
Altyn Tagh	7	8	1000-2000	>10
Kuma	5	7.5	2750	8
Changma	3	7.5	2000	5.5
Gulang	3	8	1500-2100	6.2
Haiyuan	8-9	8.5	1400	>10
Xianshuihe	10	7.5	600	3.6
Zemuhe.	3	7.5	1000	7
W. Xiaojiang	6	8	1000-2000	9-10
Qujiang	5	7.5	2000	2.7
Bengcuo.	2	8	800	7.3
Ertai F.	5	8	3000-3500	>10
Kashihe	2	8	2000-2500	4
Daqingshan	4	8	3300	2.5
Helanshan	4	8	2000-2500	1.5
Tangshan	1	7.5	7500	1.5
Huoshan	2	8	2000	8.6
Tanlu	3	8.5	3000-4000	7-9
Huashan	3	8	2000-2900	4

SUMMARY

Earthquakes are the results of lithosphere deformation. The earthquake data places important constraints on the formulating of the continental kinematic and geodynamic model. In this paper the principal features of active tectonics in Mainland China has been discussed briefly based on the seismotectonic characters of China. The main points inferred from these data are as follows:

1. The contemporary tectonic movement in Chinese Continent is driven by both the Tethy-Himalayan and Western Pacific ongoing geodynamic processes.

2. According to the space distribution and characters of seismicity, the Chinese continent can be divided into six intraplate tectonic blocks: Qinghai-Tibet Block, Gansu-Xingjiang Block, Northeast China Block, North China Block, South China Block and Southeast China Block,

bordering by two collision zones: the Himalayan Collision Zone in southwest and the East Taiwan Collision Zone in southeast.

3. The Qinghai-Tibet Block pushed by India Plate is undergoing crustal shortening, thickening, uplifting and moving northward relative to the East China resulting into the right-lateral shearing between West China and East China. Meanwhile, due to the termination of Tethys-Himalayan Mega-zone and the N-S trending asymmetrical compression, the Qinghai-Tibet Block has been extruded eastward. This dynamic process also affects much of the East China.

4. Relative motion has been found along intraplate block boundaries, but the intensity and style of the movement are considerably different from those in interplate region.

5. Four types of crustal deformation play an important role in the internal deformation inside the intraplate blocks: thrusting and shortening, conjugate strike-slip faulting, localized pull-apart extension and block rotation.

6. The earthquake recurrence interval estimated based on paleoseismic data is much longer than that in the interplate region. This implies a slow intraplate strain rate.

Acknowledgements

This study is sponsored by Chinese National Science Foundation under the Project No. 49572155. It is also partly supported by Chinese State Seismological Bureau under the Contract No. 85-07-01. The authors would like to thank Zhou Yongdong, Yang Wenlong and Zhang Hua for their assistance in the preparation of this paper.

REFERENCE

1. J.P. Avouac, P. Tapponnier, M. Bai, H. You and G. Wang. Active thrusting and folding along the northern Tien Shan and late Cenozoic rotation of the tarim relative to Dzungaria and Kazakhstan. *Jour. Geophys. Res.* 98, 6755-6804 (1993).
2. J. P. Avouac, P. Tapponnier. Kinematic model of active deformation in central Asia. *Geoph. Res. Lett.* 20, 895-898 (1993)
3. Deng Qidong, Chen Shefa, Zhao xiaolin, Tectonics,seismicity and dynamics of Longmenshan Mountains and its adjacent regions *Seismology and Geology*, Vol. 16, No. 4, 389-403(19994)
4. Ding Guoyu. The inhomogeneity of Holocene faulting, *Earthquake Res. in China*, 5, 95-105 (1991)
5. Editorial Board of Seismic Zoning Map of China, State Seismological Bureau, *Earthquake mechanism in China and its adjacent areas*. Seismological Press, Beijing China (1991) (in Chinese)
6. Editorial Board of Seismic Zoning Map of China, State Seismological Bureau, *Instruction of map of active tectonics in China and its adjacent seas*. Seismological Press, Beijing China (1991) (in Chinese)
7. P. England and P. Molnar. Right-lateral shear and rotation as the explanation for strike-slip faulting in eastern Tibet. *Lett. to Nature*, 344,140-142 (1990)
8. W.E. Holt, M. Li and A.J. Haines. Earthquake strain and instantaneous relative motions within central and eastern Asia. *Geophys. J. Int.* 122, 569-593 (1995)
9. J. G. John. Tectonics of China: continental scale cataclastic flow. Mechnical Behavior of Crustal Rocks, *Geophysical Monograph* 24, 98-105 (1981)
10. P. Molnar. The Geologic history and structure of the Himalaya. *American Scientist*, 74, 144-154 (1986)
11. P. Molnar. Continental tectonics in the aftermath of plate tectonics. *Nature*, 335, 8, 131-137 (1988)
12. G. Peltzer and P. Tapponnier. Formation and evolution of strike-slip faults, rifts, and basins during the India-Asia collision: an experimental approach. *Jour. Geophys. Res.* 93, No. B12, 15085-15117 (1988)
13. G. Peltzer, P. Tapponnier and R. Armijo. Magnitude of late Quaternary left-lateral displacements along the north edge of Tibet. *Science*, 246, 1285-1289 (1989)
14. H. Z. Wang and X. X. Mo. An outline of the tectonic evolution of China. *Episodes*, 18, nos. 1&2, 6-16

(1995)

16. X. C. Xiao and T. D. Li. Tectonic evolution and uplift of the Qinghai-Tibet Plateau. *Episodes*, 18, nos. 1&2, 31-35 (1995)

17. H. Ye, B.T. Zhang and F.Y. Mao. The Cenozoic tectonic evolution of the Great North China: two types of rifting and crustal necking in the Great North China and their tectonic implications. *Tectonophysics*, 133, 217-227 (1987)

18. H. Ye, Y.D. Zhou, Q. Zhou, W.L. Yang, G.G. Chen and C.T. Hao. Study on potential seismic sources for seismic zonation and engineering seismic hazard analysis in continental area. *IASPEI Publication Series for the IDNDR* 3, 473-478 (1993)

19. H. Ye, G.G. Chen, Q. Zhou. Study on the intraplate potential seismic sources, *Prceedings of 5th ICSZ*, Presses Academiques, 1424-1431 (1995)

20. Y.Q. Zhang, P. Vergely and J. Mercier. Active faulting in and along the Qinling Range (China) inferred from SPOT imagery analysis and extrusion tectonics of south China. *Tectonophysics*, 243, 69-95 (1995)

21. R. S. Zeng, L. P. Zhu, and et. al. A seismic source model of the large earthquakes in North China extensional basin and discussions on the genetic processes of the extensional basin earthquakes. *ACTA Geophys. Sinica*, 34, No. 3, 288-301 (1991) (in Chinese)

22. J. D. Zheng. Significance of the Altun Tagh fault of China. *Episodes*, 14, No.4,307-312 (1991)

Proc. 30ᵗʰ Int'l. Geol. Congr., Vol. 5 pp. 35-48
Ye Hong (Ed)
© VSP 1997

Crustal Dynamics in the North and East Margin of Qing-Zang (Tibet) Plateau

XIE FUREN, ZHANG SHIMIN, SHU SAIBING AND DOU SUQIN
Institute of Crustal Dynamics, SSB, Beijing, 100085, China

Abstract

The inversion of fault slip data for tectonic stress field and the analysis of crustal deformation characteristics suggest that during the period from Pliocene to early Pleistocene the north and east margin of Qing-Zang Plateau was subjected to an omnibearing compression caused by the collision of India Plate with Qing-Zang block, which resulted into mainly reverse faulting in the margin of the plateau. Since the end of early Pleistocene India Plate continues to push northward and the compressional deformation of the plateau crust increases continuously, at the same time NW-SE extension appeared on the east side of the plateau. This forms a favorable condition for the interior of plateau to slide towards east and southeast, causing the faults surrounding the plateau to change their type from thrust to strike-slip fault.

Keywords: Qing-Zang(Tibet) Plateau, tectonic stress field, crustal deformation.

INTRODUCTION

Qing-Zang (Tibet) Plateau is a result of crust shortening, thickening, and uplifting caused by the northward pushing of India Plate against the Eurasia Plate. Under the strong and continuous action of plate convergence, compressional thrust-fold mountain systems bounded by large-scale arcuate faults formed inside the plateau, while a series of active strike-slip or thrust faults of considerable scope and activeness formed surrounding the plateau, causing frequent destructive earthquakes. Therefore, the study of fault activity and tectonic deformation in the plateau margin on the crustal dynamic process of Qing-Zang Plateau is undoubtedly useful for fully understanding the evolution of Qing-Zang Plateau, It is also important for the investigation of the interaction among interior blocks of a continent.

This paper is based on previous studies [4,5,7,8,11-16]. Emphasis was placed on the evolution of tectonic stress field inverted from observed fault slip data and the analysis of neotectonic geomorphology and crustal deformation characteristics, thereby to interpret fault activities and crustal dynamics in the north and east margin of Qing-Zang Plateau.

The major active faults in the north and east margin of Qing-Zang Plateau include the following: the world-renown left-lateral strike-slip Altun fault belt in the north, Qilianshan-Hexi Corridor reverse and left-lateral strike-slip fault belt, Haiyuan-Liupanshan left-lateral strike-slip and thrust fault belt, Minjiang and Longmenshan thrust fault belts in the northeast, and Xianshuihe-Anninhe-Xiaojiang left-lateral strike-slip fault belt in the southeast. Since Quaternary most of these faults underwent a change from reverse faulting to strike slip or reverse-strike slip faulting.

EARLY QUATERNARY TECTONIC STRESS FIELD

Due to limitations of research method and observation conditions, the study of earlier tectonic stress field is still very difficult. Fortunately, by stage-divided calculations with the recently developed method of tectonic stress tensor determination from fault slip data [1,3,6], not only the present tectonic stress tensor in the study area but also the early-stage tectonic stress tensor can be obtained if observations are sufficient.

Fault movement is one of the most important manifestation of tectonic movement of the crust, it is the product of direct action of tectonic stress on crustal rocks. A great amount of fault surface structures (striation, steps, etc.) existing in rocks faithfully recorded the directional characteristics of slips caused by tectonic stress on the fault, therefore they must contain the information of tectonic stress field which caused fault movement. The stress tensor inversion method using fault slip data is to determine the tectonic stress state in the fault area based on a set of observation data which contains the characteristics of fault movement. The essence of the method is to fit the calculated shear stress direction to the slip direction on the fault surface, then finally yield 4 characteristic parameters of the stress tensor, that is, directions of 3 principal stresses and a stress ratio $R=(s_2-s_3)/(s_1-s_3)$ which represents the proportional relation of principal stress values.

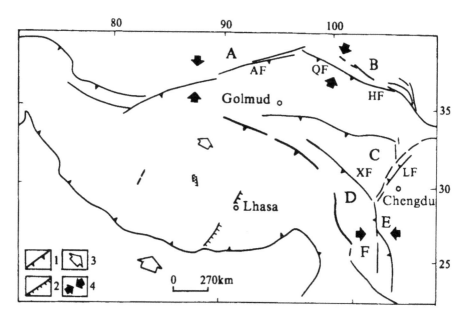

Fig. 1 Map of early Quaternary tectonics stress field in the north and east marginal region of Qing-Zang Plateau. 1.reverse fault; 2.normal fault; 3.motion direction of block; 4.principal compressional stress direction. AF:Altun fault; QF:Qilianshan fault; HF:Haiyan-Liupanshan fault; LF:Longmenshan fault; XF:Xianshuihe-Anninhe-Xiaojiang fault.

The method of stage-divided stress calculation is based on the following argument: among the N observational data if n (n<N) were produced in the same tectonic stress stage, then the fitting error of these observed striation to the shear stress direction of corresponding stress tensor on the fault surface should be much smaller. Thus by repeated trial and adjustment on a computer a reasonable stress stage division and corresponding fault movement

combination can be finally determined [2,17].

Table 1. Characteristics of early Quaternary tectonic stress tensors in the north and east borders of Tibet Plateau

Division	Location of site	σ_1		σ_2		σ_3		R	Stress regime
		Azimuth (°)	Dip (°)	Azimuth (°)	Dip (°)	Azimuth (°)	Dip (°)		
A	Gasi	1	38	270	1	180	52	0.02	reverse
	Kulesayi	353	37	188	52	89	7	0.64	strike–slip
	Yitunbulake	10	28	123	36	253	42	0.16	reverse
	Dachonggou	168	10	258	5	4	79	0.77	reverse
	Changcaogou	358	27	94	10	202	60	0.16	reverse
B	Kushuigou	215	4	124	11	326	78	0.46	reverse
	Heishuihekou	37	7	305	17	150	72	0.66	reverse
	Shenjiazhuang	65	17	326	29	182	56	0.45	reverse
	Jiangou	235	3	325	7	125	82	0.26	reverse
	Yehupo	237	22	136	24	4	56	0.48	reverse
	Liupanshan	75	1	345	1	213	88	0.48	reverse
D	Dandou	248	49	93	38	353	13	0.66	normal
	Upper,Down yajialuo	237	1	327	10	142	80	0.09	reverse
	Songlin pass	74	21	344	1	250	69	0.39	reverse
	Western Litang	20	11	131	62	284	26	0.53	strike–slip
	Jiawa	18	26	228	61	114	13	0.66	strike–slip
E	Xichang	253	4	137	82	344	7	0.39	strike–slip
	Lugu	66	23	293	57	165	21	0.64	strike–slip
	Southern Mianning	57	17	154	21	292	62	0.61	reverse
F	Jingjiang	89	17	234	69	356	11	0.48	strike–slip
	Yongsheng	95	25	271	65	5	2	0.09	strike–slip
	Zhongdian	100	40	230	38	344	28	0.69	normal
	Lijiang	256	32	127	44	6	28	0.67	strike–slip
	Jianchuan	75	14	180	47	333	40	0.38	strike–slip
	Shaxi	75	6	184	73	341	16	0.50	strike–slip
	Eryuan	274	31	43	46	166	27	0.65	strike–slip
	Shaping	113	1	19	77	203	13	0.44	strike–slip
	Xiaguan	273	56	57	28	156	17	0.75	normal
	Puer	97	30	317	53	199	19	0.30	strike–slip

By using this stage-divided calculation method of tectonic stress we obtained an outline map of early Quaternary tectonic stress field in the north and east marginal region of Qing-Zang Plateau (Table 1, Fig.1). Due to limitations of observation the data for early Quaternary stress tensors were not sufficient, therefore the result is only a general outline of the early stage stress field. It can be seen from the result that in the Altun fault zone at the north boundary of the plateau the early Quaternary tectonic stress field was mainly a compression in near-NS direction, the stress regime was of reverse faulting type; in the northeast marginal region of the plateau the principal compressional stress of the early Quaternary tectonic stress field was in northeast direction, the stress tensor was mainly of reverse faulting type; in the northern part of east margin of the plateau the principal compression of early Quaternary tectonic stress field was in NE or NEE direction, in its southern part in west Sichuan and Yunnan region the principal stress direction was NEE or near-EW, the tectonic stress tensors were mostly of reverse faulting type.

According to tectonic deformation and change of faults movement on the north and east margin of Qing-Zang Plateau, the upper time limit of this stress field stage was determined to be from the end of early Pleistocene to middle Pleistocene [16].

RECENT TECTONIC STRESS FIELD

On the contemporary tectonic stress field of Qing-Zang Plateau there have been many studies based on focal mechanism solutions, numerical simulations and other methods [9,18,19]. In the recent years we studied the resent tectonic stress field with fault slip data, our results were basically similar to that from focal mechanism solutions and some stress measurements (Table 2, Fig.2), which indicates that the recent tectonic stress field in the north and east margin of Qing-Zang Plateau is continuous and stable in a considerably long geological time period.

The basic features of the recent tectonic stress field in the north andeast margin of Qing-Zang Plateau inverted from active fault slip data are as follows. In the Altun fault zone in north (region A) the maximum compressional stress direction from west to east changes from NNE to NE, the stress tensor is of reverse or strike-slip type (Fig.2, Table 2); In Qilianshan-Hexi Corridor fault belt to Haiyuan-Liupanshan fault belt region (region B) in northeast of the plateau the maximum compressional stress direction from west to east changesfrom NE or NEE to near-EW, the stress tensor in the west segment of Qilianshan- Hexi Corridor fault belt is mainly of reverse faulting type, in the vast area from east segment to Haiyuan fault belt it is of strike slip type (Fig.2, Table 2). In Minjiang fault and Longmenshan fault belt region (region C) in the northern part of east margin of Qing-Zang Plateau, the maximum compressional stress direction is NEE or near-EW, the stress tensor in Longmenshan area is of reverse faulting type, while in Mingjiang fault belt to the west it is of strike-slip type (Fig.2, Table 2). In Anninhe-Xiaojiang fault belt area (region E) in the southern part of east margin the maximum compressional stress is in NW-SE direction (Fig.2, Table 2); In Xianshuihe fault zone (region D) west of Anninhe-Xiaojiang fault belt the maximum compressional stress direction is near east-west, the tectonic stress regime is of strike-slip type (Fig.2, Table 2). In the northwestern area in Yunnan the maximum compressional stressis in NNW-SSE direction, the tectonic stress regime is of strikeslip type (Fig.2, Table2).

Synthesizing the tectonic stress field determination results from fault slip data, focal mechanism solutions, and in-situ stress measurements, the following basic characteristics of the recent tectonic stress field in the north and east margin of Qing-Zang Plateau can be obtained:

(1) The action of recent tectonic stress field is mainly a horizontal compression. The great majority of maximum principal compression axes determined from focal mechanism

solutions and fault slip data are nearly horizontal, the stress regimes are mostly of strike-slip type (Table 3). The horizontal component of fauting activities is clearly greater than the vertical component.

Table 2. Characteristics of recent tectonic stress tensors in the north and east borders of Tibet Plateau

Division	Location of site	σ_1		σ_2		σ_3		R	Stress regime
		Azimuth (°)	Dip (°)	Azimuth (°)	Dip (°)	Azimuth (°)	Dip (°)		
A	Yuohepeyisayi	197	5	302	72	105	17		strike–slip
	QigeQuangou	54	10	146	7	269	77	0.37	reverse
	Gebiling	230	10	140	1	46	80	0.28	reverse
	Lapeiquan	50	11	141	8	268	76	0.40	reverse
	Yemaquan	43	28	144	21	267	54	0.08	reverse
	Annanba	72	20	285	67	166	11	0.09	strike–slip
	Mobar	210	5	77	83	300	5	0.84	strike–slip
	Qingyazi	55	20	300	40	168	39	0.15	strike–slip
	Changcaogou	49	16	316	9	197	71	0.28	reverse
B	Sangequan	238	13	63	77	328	1	0.35	strike–slip
	Huangjiagou	60	15	150	3	251	74	0.45	reverse
	Fodongmiao	62	10	329	20	177	68	0.57	reverse
	Jiayuguan	33	4	302	11	145	78	0.42	reverse
	Zhuozishan	67	8	160	18	314	70	0.62	reverse
	Heishuihekou	225	2	121	81	315	9	0.74	strike–slip
	Shenjiazhuang	250	25	111	58	349	18	0.43	strike–slip
	Huangliangtan	242	2	151	50	333	40	0.71	strike–slip
	Gaowanzi	63	1	155	87	333	4	0.89	strike–slip
	Jiangou	75	5	175	62	342	28	0.21	strike–slip
	Youfangyuan	85	10	216	75	353	11	0.29	strike–slip
	Caixiangbao	272	9	88	81	182	7	0.42	strike–slip
	Xixia Reservoir	85	11	331	65	180	22	0.46	strike–slip
	Heshangpu	267	1	1	70	177	20	0.60	strike–slip
C	Xuebaoding	243	38	82	50	340	9	0.48	strike–slip
	Zhangla pass	83	9	210	75	351	12	0.31	strike–slip
	Chuanzhusi	266	8	142	75	358	12	0.30	strike–slip
	Suosuozhai	76	23	312	53	179	27	0.65	strike–slip
	Jiaochang	284	32	120	57	19	7	0.44	strike–slip
	Guanxian	75	9	166	11	304	76	0.39	reverse
	Heishuihekou	50	1	319	28	142	62	0.47	reverse

(Continued of Tbale2.)

Division	Location of site	σ_1 Azimuth (°)	σ_1 Dip (°)	σ_2 Azimuth (°)	σ_2 Dip (°)	σ_3 Azimuth (°)	σ_3 Dip (°)	R	Stress regime
D	Zuwo	74	2	179	82	343	3	0.44	strike–slip
	Dandou	273	11	18	51	175	37	0.39	strike–slip
	Xialatuo	88	27	220	53	345	24	0.58	strike–slip
	Qiajiao	276	8	16	53	180	36	0.42	strike–slip
	Mazi	110	18	232	59	12	24	0.34	strike–slip
	Upper,Down Yajialuo	99	25	241	59	0	17	0.48	strike–slip
	Northern Qianning	96	9	354	53	193	36	0.25	strike–slip
	Zheduotang	290	13	122	76	21	3	0.87	strike–slip
	Southern Kangding	100	5	343	80	191	9	0.43	strike–slip
	Western Litang	274	25	156	46	22	33	0.24	strike–slip
	Jiawa	98	32	289	58	191	5	0.54	strike–slip
E	Southern Mianning	112	10	270	79	21	4	0.63	strike–slip
	Xichang	307	3	40	37	213	53	0.40	reverse
	Wudaoqing	154	22	18	61	251	18	0.90	strike–slip
	Mabian	117	7	24	21	224	67	0.64	reverse
	Jingjiang	167	18	301	64	71	17	0.46	strike–slip
	Yongsheng	350	34	206	50	93	18	0.57	strike–slip
	Qiaojia,Dongchuan	277	15	158	61	14	24	0.46	strike–slip
	Songming	118	9	252	77	27	9	0.95	strike–slip
	Jiangchuan	292	27	61	52	188	26	0.93	strike–slip
	Tonghaixian	338	2	71	56	247	34	0.45	strike–slip
F	Zhongdian, Hutiaogorge	350	24	144	64	255	10	0.50	strike–slip
	Lijiang	23	12	140	64	288	32	0.48	strike–slip
	Jianchuan	199	9	303	57	104	31	0.40	strike–slip
	Northern Shaxi	166	8	271	62	73	26	0.54	strike–slip
	Eryuan	200	8	95	63	294	26	0.35	strike–slip
	Shaping	203	13	33	77	293	2	0.44	strike–slip
	DingXiling	195	3	287	34	101	56	0.60	reverse
	Xiaguan	173	12	62	60	269	27	0.75	strike–slip
	Nanjian	165	4	257	30	68	60	0.71	reverse
	Lincang	2	37	193	53	96	5	0.59	strike–slip
	Puer	197	6	88	71	289	18	0.46	strike–slip
	Yuanjiangxian	165	16	3	73	256	5	0.31	strike–slip

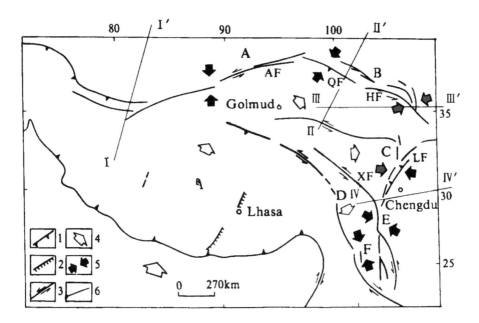

Fig. 2 Map of recent tectonic stress field in the north and east marginal region of Qing-Zang Plateau. 1.reverse fault; 2.normal fault; 3.strike-slip fault; 4.motion direction of block; 5.principal compressinal stress direction; 6.profile line.

Table 3. Characteristics of recent tetonic stress field in the east border of Tibet Plateau

Characteristic parameter	Seismic source mechanism solution	Results determined from data on fault sliding
Average elevation angle (dip) of $P(\sigma_1)$axes(°)	15.9°	(14.8)
Average elevation angle (dip) of $P(\sigma_3)$axes(°)	15.7°	(18.1)
Percentage of strike—slip stress structure	88%	88%
Percentage of reverse stress structure	6%	12%
Percentage of normal stress structure	6%	——
Average dip of sections (°)	73.7°	——
Average dip of fault planes (°)	——	67.6
Average pich of striae (°)	——	29.8

[*] After KAN et al.,1977

(2) The spatial variation of recent tectonic stress field is concordant. Although the recent tectonic stress field in the north and east margin of Qing-Zang Plateau shows complicated variations, the general tendency of variation is rather concordant, the maximum principal compression direction from northwest to southeast rotates gradually from NNE to SSE.

(3) The recent tectonic stress field shows regionalized characteristics. These characteristics are expressed not only in different stress directions, but also in different stress tensor structures (Table 4). Controlled by faulting structures and dynamic conditions, they are

mainly divided into two stress regions of strike-slip faulting type and reverse faulting type.

Table 4. Division of recent tectonic stress in the north and east borders of Tibet Plateau

No.	Division	Principal compressive stresS (σ_1)		Stress regime
		Azimuth	Dip	
A	Altun	NNE, NE	Near horizontal	strike–slip
B	Qilianshan —Liupanshan	NE, NEE, sub–EW	Near horizontal	reverse, strike–slip
C	Minjiang, Longmenshan	NEE, sub–EW	Near horizontal	strike–slip revers
D	Xianshuihe, Litang	sub–EW	Near horizontal	strike–slip
E	Anninghe— Xiaojiang	SE	Near horizontal	strike–slip
F	NW Yunnan	SSE	Near horizontal	strike–slip

(4) The principal compressional stress direction of recent tectonic stress field rotated clockwise with respect to the early Quaternary stress field.

CHARACTERISTICS OF CRUSTAL DEFORMATION

The dynamical process in the crust is expressed and preserved by deformation. Therefore, the study of crustal deformation characteristics is an effective way to investigate the tempo-spatial evolution pattern of crustal dynamics.

The uplifting of Qing-Zang Plateau can be divided into an earlier relatively stable stage (Paleocene to early Miocene) and a later rapid uplifting stage (since middle late Miocene) [10].

Relatively stable stage
From Paleocene to early Miocene the Qing-Zang block emerged from the sea surface and rose to about 1200-1500m ASL, which took a long time of more than 40 million years.

In this period the tectonic movement was relatively stable, geomorphically it experienced a long and extensive peneplanation process [20]. A gently rolling peneplain of several hundreds to about 1500m above sea level was formed across the Qing-Zang, Tarim, Alashan, Ordos, and South China blocks. The related sediments are preserved in slowly depressed areas (e.g. some peripheral regions surrounding the plateau), they mainly consist of fine-grained materials with particle size of clay, silt, and fine sand.

The crust overlies on a relatively homogeneous and rheologic mantle. Therefore in the process of peneplanation the crust inevitably underwent through a gravitational isostatic adjustment. Since the land surface eventually formed an even peneplain surface, it is inferred that the gravitational isostasy is relatively complete. Assuming that the crustal density is homogeneous, then isostasy must result in a homogeneous crustal thickness. This homogeneous crust is tentatively named as the datum crust in order to be distinguished from the crust deformed after Miocene.

Because of the incompleteness in the actual peneplanation process, the altitude of the peneplain surface is slightly higher at the center of the continent than at its margin. This directly affects the degree of isostatic compensation of the crust. Therefore the thickness of the datum crust is also slightly greater in the continent center. The northern part of Alashan block and the Ordos block are not significantly deformed since the Cenozoic, their crustal thickness is 42-44km, which can be taken as the datum crust thickness there. The South China block is close to the ocean, its central part is not obviously deformed since Cenozoic, the crustal thickness is about 40km, which can be taken as the datum crustal thickness there.

Rapid uplifting stage
From middle Miocene to Quaternary is the rapid uplifting stage of the Qing-Zang Plateau. The plateau surface rose from 1200m ASL to over 5000m in less than 20 million years.

This stage was characterized by strong tectonic deformation, volcanism and even metamorphism. The peneplain surface formed in early Tertiary was broken, coarse-grained materials such as medium-coarse sand and gravel were widely distributed in the sediments above middle Miocene series, indicating that the relief was already obviously rough. Strong crustal deformation such as shortening, thickening, folding, and faulting also took place in Qing-Zang block and surrounding areas.

Crust deformation characteristics at the north margin of Qing-Zang Plateau
West segment
The crust of Qing-Zang block on the south side of the margin has thickened by 16-24km; the crust of Tarim block on the north side thickened by 4-12km; to further north the crust of Tianshan block thickened by 10km (Fig. 3).

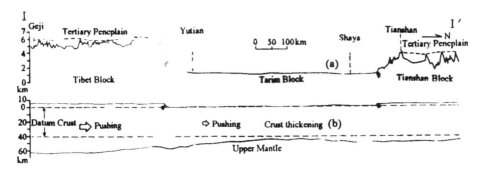

Fig. 3   Geji-Shaga (a)topographic profile (b)crust morphologic profile.

East segment
The crust of Qing-Zang block on the south side of the margin has thickened by 14-18km; the crust of the southern end of Alashan block on the north side has thickened by 12km. Towards north up to the boundary between China and Mongolia the crust is gradually transformed into the datum crust(Fig. 4).

Crustal deformation characteristics at the east margin of Qing-Zang Plateau
North segment

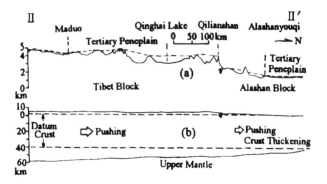

Fig. 4   Maduo-Alashanyouqi (a)topographic profile (b)crust morphologic profile.

The crust of Qing-Zang block on the west side of the margin has thickened by 14-16km, near to the block margin the crust thins rapidly to the datum crustal thickness. In Ordos block on the east side, the crust at the west end is 4km thinner than the datum crustal thickness, towards east it is gradually transformed into the datum crust(Fig. 5).

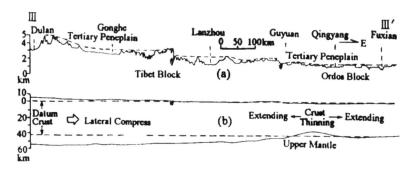

Fig. 5   Dulan-Fuxian (a)topographic profile (b)crust morphologic profile.

South segment
On the west side the crust of Qing-Zang block has thickened by 16-20km, near to the block margin the crust thins rapidly. On the east side the crust at the west end of South China block is 2km thinner than the datum crustal thickness, to the east it is gradually transformed into the datum crust (Fig. 6).

DYNAMIC EVOLUTION MECHANISM

The effect of collision between India and Eurasia Plates on the interior of Qing-Zang Plateau involves longitudinal compression as well as lateral squeezing. This squeezing is most distinctive in the region between India Plate and Tarim block, which resembles the sideway squeezing out of relatively plastic material between two rigid blocks. This lateral flow of materials caused the disintegration and transportation of the interior and marginal blocks of Qing-Zang block.

The analysis of crustal tectonic deformation shows that crust thickening is seen in the vast area on both sides of the northern margin of Qing-Zang Plateau, indicating that the pushing action of India Plate on Qing-Zang block is transmitted through the north margin to further north. Whereas at the east margin of Qing-Zang Plateau crustal thickening rapidly decays here, to its east the crustal thickness is even thinner than the datum crust. This suggests that the eastward lateral pushing action of India Plate ends at the east margin of Qing-Zang Plateau. On the other hand the northward movement of Qing-Zang block produces a northwesterly traction to the east-side block, making the crust at the west end of the east-side South China block thin locally. It was this local extension of the west end of South China block that provided a favorable condition for the southeastward flow of the eastern part of Qing-Zang Plateau in the later stage.

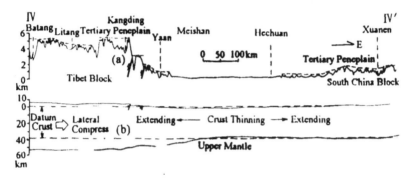

Fig. 6 Batang-Xuanen (a)topographic profile (b)crust morphologic profile.

The evolution of tectonic stress field can quite well explain the process of this crustal dynamic evolution. In the early stage of neotectonic movement, i.e., before the early Quaternary, the north and east margin was mainly under the frontal and lateral compression caused by the convergence of India and Eurasia Plates. In this period India Plate pushed frontally the Qing-Zang block to the north, at the same time caused intense lateral compression in its northeast. In Burma and west Yunnan region east of Asam wedge-shaped block, this lateral pressure was expressed by strong compression in NEE to near-EW direction, giving rise to a stress field with near-EW principal compression in this region (Fig.7A), and produced a series of closely bundled structures and compressional belts as represented by the transverse mountain ranges. In the Sichuan-Tibet region northeast of Asam wedge, there was NNE or NE compression.

During the period of early to middle Pleistocene India Plate continued to push Eurasia continent to the north. After suffering a great amount of longitudinal compressional deformation in a long time, the Qing-Zang block in between India Plate and the rigid Tarim block started a large-scale lateral transportation of material. Induced by crustal thinning and extensional environment on theeast side, the east margin of the plateau started to disintegrate and move towards east and southeast, the stress field was readjusted. Owing to the lateral movement of the northeast marginal block of the plateau, the lateral pressure caused by the frontal action of India Plate was alleviated, thus making the lateral pressure of India Plate on southwest China region become secondary. Thereafter the stress field in southwest China in the latter neotectonic stage was formed (Fig.7B). This stress pattern was formed in early to middle Pleistocene and lasted to the present time.

CONCLUSION

(1) Since the early neotectonic stage the north and east margin of Qing- Zang Plateau experienced two major stages of tectonic stress field. The change and adjustment between these two tectonic stress fields took place approximately in early to middle Pleistocene. The first stage lasted approximately from Pliocene to early Pleistocene, when the tectonic stress field was featured by a maximum principal compression perpendicular to the boundary of the Plateau, and was basically of reverse faulting type. The second stage is from the end of early Pleistocene to present, that is the recent tectonic stress field. The direction of maximum compressional stress rotated clockwise with respect to the previous tectonic stress field, the stress regime was mainly of strike-slip type.

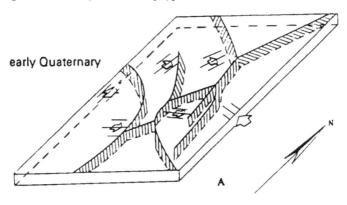

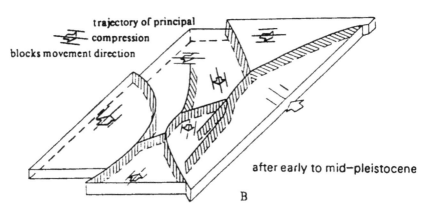

Fig. 7 Evolution pattern of tectonic stress field during Quaternary in the east margin of Qing-Zang (Tibet) Plateau.

(2) The deformation of crust of Qing-Zang Plateau and surrounding regions underwent through two major stages. In the first stage from Paleocene to early Miocene, the surface of crust was subjected to peneplanation of long time, resulting in a peneplain surface about 1200-1500m above sea level. Owing to gravitational isostatic adjustment a datum crust with relatively homogeneous thickness was formed (the datum crustal thickness was 42-44km in the north margin, about 40km in the southeast margin). In the second stage from late Tertiary to Quaternary, especially since Pliocene, the Qing-Zang Plateau rose rapidly from about 1500m ASL to above 5000m. The crust of its northern part and outside region thickened by

16-24km and 4-12km respectively. In the southeast of the Plateau the crust thickened by 16-20km, while that outside the periphery thinned by 2-4km.

(3) The phenomena of crustal thickening or thinning at the north and east margin of Qing-Zang Plateau indicate that the compression of India Plate against Qing-Zang block is transmitted through the northern margin of Qing-Zang block to further north. While at the southeast margin of the block NW-SE extension appears on its east side due to large northward movement of Qing-Zang block.

(4) The crustal dynamic process at the north and east margin of Qing-Zang Plateau can be reasonably explained by the evolution of tectonic stress field. In the early stage strong lateral compression was produced on the east side of the plateau. In the latter stage, after suffering a great amount of longitudinal compressional deformation in a long time, the Qing-Zang block started a large-scale lateral transportation of material. Induced by crustal thinning and extensional environment, block disintegration and migration towards east and southeast took place at the east margin of the plateau, Consequently, the stress field has been re-adjusted.

REFERENCES

1. J. Angelier, Determination of the mean principal direction of stresses for a given fault population, *Tectonophysics*, 56, 17-26.(1979)
2. J. Angelier, Tectonic analysis of fault slip data sets, *J. Geophys. Res.*, 89, B7, 5835-5848.(1984)
3. E. Carey, and B. Brunier, Analyse theorique et numerique d' un modele ecanique elementaire applique a l' etude d' une population de failles. *C. R. Acad. Sci.*, 891-894. Ser. D, 279, Paris.(1974)
4. Ding Guoyu and Lu Yanchou, A preliminary disussion on the status of recent intraplate motions in China, *Science Bulletin*, 33,51, 52-57.(1986)
5. Ding Guoyu, Some problems on active tectonics in Qinghai-Tibet Plateau, *Northwestern Seismology Journal*, 10, Suppl., 1-11(in Chinese).(1988)
6. A. Etchecopar, G. Vasseur. and M. Daignieres, An inverse problem in microtectonics for the determination of stress tensors from fault striation analysis. *J. Struct. Geol.*, 3, 1, 51-55. (1981)
7. Institute of Geology, SSB and Seismological Bureau of Ningxia, *Haiyuan Active Fault Zone*, Seismological Press, Beijing (in Chinese).(1990)
8. Institute of Geology, SSB and Lanzhou Institute of Seismology, SSB, *Qilianshan-Hexi Corridor Active Fault System*, Seismological Press, Beijing (in Chinese). (1993)
9. Kan Rongju, Zhang Sichang, Yan Fengtong and Yu Linsheng, Present tectonic stress field and its relation to the chracteristics of recent tectonic activity in southwestern China. *Acta Geophysica*, 20, 1, 96-108 (in Chinese).(1977)
10. Li Bingyuan et al., *Quaternary Geology in Xijang* (Tibet), Science Press, Beijing(in Chinese).(1983)
11. P. Molnar, and P. Tapponnier, Cenozoic tectonics of Asia: effects of a continental collision. *Science*, 189, 4201, 419-425.(1975)
12. P. Molnar, and H. Lyon-Caen, Fault plane solution of earthquakes and active tectonics of the Tibetan Plateau and its margins, *Geophy. J. Int.*, 99, 123-153. (1989)
13. P. Tapponnier, G. Paltzer, and R. Armijo, On the mechanics of the collision between India and Asia, in *Collision Tectonics*, edited by M.P. Coward and A.C. Reis, Geol. Soc. London Spec. Publ., 19, 115-157. (1986)
14. Working group of Altun active fault belt, SSB, *Altun Active Fault Zone*, Seismological Press, Beijing(in Chinese). (1992)
15. Xie Furen and Liu Guangxun, Neotectonic stress field in the central segment of Altun fault zone, China, determined from fault striations, *Earthquake Research in China*, 6, 1, 99-112. (1992)
16. Xie Furen, Zhu Jingzhong, Liang Haiqing, and Liu Guangxun, Basic characteristics of the recent tectonic stress field in southwest China, *Acta Seismologica Sinica*, 6, 4, 843-855. (1993)
17. Xie Furen, Zhu Jingzhong, and Shu Saibing, Division of stages of Quaternary tectonic stress field in Xianshuihe fault region, *Seismology and Geology*, 17, 1, 35-43(in Chinese). (1995)
18. Xu Zhonghuai, Wang Suyun, Huang Yurui, Gao Ajia, Jin Xiaofeng and Chang Xiangdong, Dsirections of mean stress axes in southwestern China deduced from microearthquake data. *Acta Geophysica Sinica*, 30, 476-486 (in Chinese).   (1987)
19. Yan Jiaquan, Shi Zhenliang, Wang Suyun and Huan Wenlin, Some features of the recent tectonic stress field

of China and environs, *Acta Seismologica Sinica*, 1, 9-24 (in Chinese). (1979)

20. Yang Yichou et al., *Geomorphology of Xizang* (Tibet), Science Press, Beijing(in Chinese). (1983)

Proc. 30th Int'l. Geol. Congr., Vol. 5 pp. 49-61
Ye Hong (Ed)
© VSP 1997

Evidence of Neotectonic Activity in the South Carolina Coastal Plain.

PRADEEP TALWANI AND RONALD T. MARPLE
Department of Geological Sciences, University of South Carolina, Columbia, SC 29208, U.S.A.

Abstract

The Coastal Plain of South Carolina was the location of the largest historical earthquake along the East Coast of the USA. Due to the low level seismicity and thick sediment cover the location and nature of the causative fault is not well understood. By integrating current seismicity, river morphological, geomorphological, paleoseismic and geophysical data we have discovered an ~200 km long NNE-trending fault zone buried below the Coastal Plain of South Carolina. It lies along a zone of river anomalies and the Woodstock fault. Its length is adequate to generate the M_w 7.3, 1886 Charleston earthquake. Other corroborative data indicate that there has been tectonic activity on this fault for at least the last one million years. Currently the most seismically active part of this fault zone is its southern end where it intersects with the NW-trending Ashley River fault zone. The intersection provides a location for stress accumulation due to plate tectonic forces and pursuant seismicity.

Keywords: Neotectonics, River Morphology, SE US Atlantic Coastal Plain, 1886 Charleston earthquake

INTRODUCTION

The August 1886 Charleston, South Carolina earthquake was the largest earthquake to hit the eastern United States in historical times. The exact nature and cause of this M_w 7.3 [7] earthquake and current seismicity near Charleston have been the subject of several multidisciplinary studies in the last two decades [5, 17]. The meizoseismal area lies in an intraplate region characterized by low strain rates and long return periods. It is also in a region of heavy vegetation and swamps with the brittle faults covered by hundreds of meters of sediments. In such a location obvious surficial manifestations of long faults, like fault scarps are absent. To delineate such features we have to rely on indirect lines of evidence. In this study we present a brief description of the integration of a variety of data used to a) delineate a major buried structure under the Coastal Plain of South Carolina and b) present evidence of neotectonic activity on it. The detailed data are presented elsewhere [11]. The data are used to delineate the buried fault, which is the southern segment of a major fault system under the Coastal Plain extending from South Carolina to Virginia, which we have labeled the East Coast Fault System (ECFS) [11]. These data consist of the observations following the 1886 earthquake; current seismicity; zone of river anomalies, (ZRA) [10]; shallow stratigraphic, seismic reflection and other corroborative data. Evidence of neotectonic activity on the ECFS was obtained from a variety of data, including current seismicity, river morphology, releveling profiles, and shallow stratigraphic and paleoseismological data. Integration of these data suggest the presence of a buried N15°E-trending fault zone ~200 km in length, with ongoing seismicity near its southern end. This fault was the probable cause of the 1886 earthquake and possibly six other large earthquakes in the past 5000 years.

DATA SUGGESTING THE PRESENCE OF A MAJOR BURIED NNE TRENDING STRUCTURE

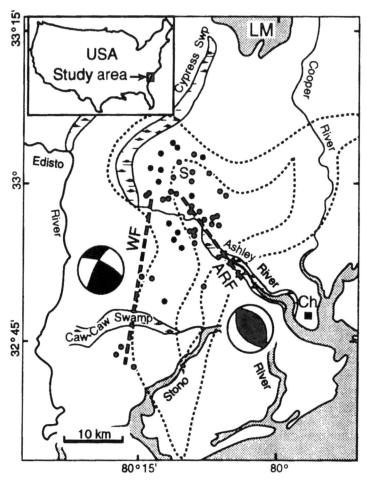

Fig. 1 Locations of seismicity (1974-1996, solid dots) compared with the zone of river anomalies (ZRA, NNE-trending striped pattern), Ashley River and Woodstock faults (ARF and WF, respectively) (dashed lines, from [20]), and Sloan's isoseismals (dashed closed contours near Summerville-S, after [4]). Focal mechanisms for ARF and WF taken from [8]. Ch-Charleston, LM-Lake Moultrie.

Seismicity

Talwani [20] suggested that the seismicity in the Charleston area was occurring at the intersection of two buried faults-the NNE-SSW trending Woodstock fault and the NW-SE trending Ashley River fault (ARF). The Woodstock fault is associated with right lateral strike-slip faulting whereas the ARF is associated with reverse faulting. Additional seismicity data through 1990[8] further supported Talwani's original conclusions. The sense of faulting was obtained from composite and single event fault plane solutions (Fig. 1). Instrumental seismicity (1980-1996) consists of a dense cluster near the inferred intersection of the two fault zones, although some evidence of the two fault trends can be seen (Fig. 1). Further evidence of two or more faults can be seen in the three lobes of the isoseismals in the meizoseismal area of the 1886 earthquake (Fig. 1) [4]. The southern lobe is roughly parallel to the inferred buried Woodstock fault and suggests a causative association.

Zone of River Anomalies
The Woodstock fault may be a part of a longer buried fault system associated with an ~200 km long, 10-15 km wide zone of river anomalies (ZRA) [10]. The ZRA trends ~N15°-20°E through the outer and middle Coastal Plain provinces of South Carolina from just south of Summerville to just east of Florence (Fig. 2). It is defined by the systematic, collinear alignment of NNE-convex arc shaped segments in river valleys. The ZRA is characterized by upwarped floodplains, changes in channel sinuosity, increased channel incision and floodplain dissection, changes from meandering to anastomosing river patterns and changes from generally symmetric to asymmetric cross-valley shapes [9, 10].

South of Summerville and near the main area of microseismicity, an ~10 km long reach of the Ashley River displays increased incision where it traverses the ZRA (Fig. 3). The floodplain downstream from the ZRA is a swampy salt marsh whereas to the west, where it traverses the ZRA, it is dry, wooded and elevated 2 to 3 m above the average water level of the channel. Upstream from the ZRA the floodplain changes back to a swampy environment, but with an anastomosing pattern (Cypress Swamp, Fig. 3).

The next three rivers to the north, the Santee, Black and Lynches, display conspicuous, NNE-convex, arc-shaped curves in their valleys where they traverse the ZRA (Fig. 2). Evidence of tectonic uplift inferred from the morphology of these rivers is discussed in a later section.

Shallow Stratigraphy
Auger-hole and well data along the ZRA to the north and south of Lake Moultrie reveal uplifted stratigraphy. Investigation of the anomalously oriented, early Pleistocene Summerville barrier and underlying shallow marine sediments near Summerville reveals that they were deposited on a NNE-trending, buried structural high in the pre-Plio-Pleistocene surface near Summerville [23].

In northeastern South Carolina, the base of a prominent, widespread clay unit within the Black Creek Formation (Upper Cretaceous) is upwarped ~45 m beneath the northern end of the ZRA between the Lynches and Pee Dee rivers. Further south along the east side of the ZRA between the Lynches and Santee rivers, this Upper Cretaceous horizon exhibits a west-side-up flexure, which suggests faulting or folding of this horizon. The contours in the southwestern part of the map area are poorly constrained due to the lack of subsurface data in this area. This linear, NNE-trending area of upwarped Upper Cretaceous sediments between the Santee and Pee Dee rivers is aligned with the inferred uplift associated with the Summerville barrier to the south.

Geophysical Data
Aeromagnetic data of the South Carolina Coastal Plain that were enhanced and filtered [14] to display subtle northeast-southwest oriented magnetic gradients revealed an ~N20°E trending, ~40 km long linear aeromagnetic anomaly near Summerville. This magnetic anomaly is approximately centered on a linear topographic high, Summerville barrier and ZRA (Fig. 3). South of Summerville an ~20 km wide area of recent uplift was detected along a releveling profile (line 9 in [15]). It coincides with the southern end of the ZRA and surface

projection of the Woodstock fault of Talwani [20]. These features are parallel to and along
the west side of the main axis of the highest intensity isoseismals of the 1886 Charleston,
South Carolina earthquake (Figs. 1 and 3).

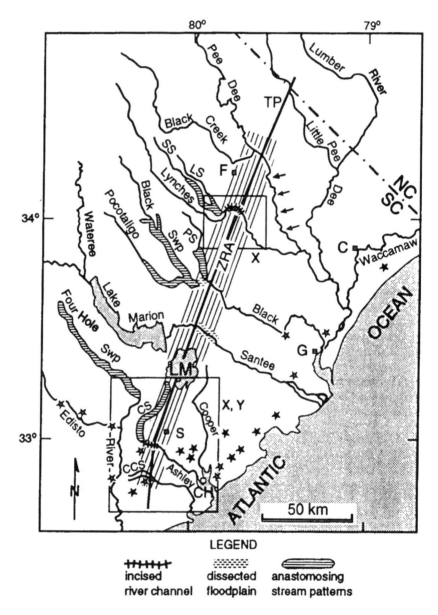

LEGEND

┼┼┼┼┼┼┼   ░░░░░░░   ⬭
incised   dissected   anastomosing
river channel   floodplain   stream patterns

Fig. 2 Rivers of South Carolina Coastal Plain with zone of river anomalies (ZRA, striped area), anastomosing
stream patterns, pre-1886 sandblow sites and topographic profile (bold line denoted TP) approximately along axis
of ZRA. The arrows along the north side of the Pee Dee River and downstream of the ZRA denote portion of the
river flowing against the southwest side of its valley. C-Conway, CCS-Caw Caw Swamp, CH-Charleston, CS-
Cypress Swamp, F-Florence, G-Georgetown, LM-Lake Moultrie, LS-Lake Swamp, PS-Pudding Swamp, S-
Summerville, SS-Sparrow Swamp. Locations of Figs. 1, 3, and 4 are also shown.

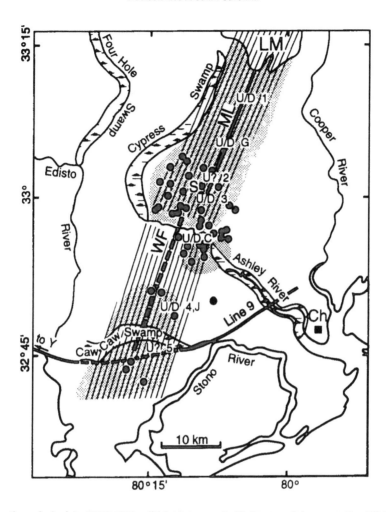

Fig. 3 Locations of seismicity (1980-1996, solid dots) compared with the zone of river anomalies (ZRA, NNE-trending striped pattern) and locations of various geological features. Woodstock fault (dashed line denoting WF taken from [21]). Shaded areas denote topographically high areas inferred from topographic profiles (see Fig. 4 of [10]). Line 9 shows part of leveling line from Yemassee (Y) to Charleston (Ch) east of the Edisto River (from [15]). The uplifted part of Line 9 is dashed. Buried faults and areas of upwarped sediments inferred from seismic reflection data are denoted by U/D (U on relative upthrown side) and U?, respectively. Locations 1-5 are taken from [11]). J-western edge of missing 'J' horizon, C-Cooke fault, G-Gants fault [6]. Ch-Charleston, LM-Lake Moultrie, ML-linear aeromagnetic anomaly from [14], S-Summerville.

Shallow seismic reflection data are available along certain portions of the ZRA (see [11] for details). The EXXON Exploration Company acquired a 55 km long seismic reflection profile across the South Carolina Coastal Plain (unpublished data) during the mid 1980s that traverses the ZRA between the Black and Lynches rivers. These data revealed two steeply dipping faults about 3.8 km apart that are approximately centered on the ZRA. Displacements along the steep, west dipping fault toward the east side of the ZRA decrease from about 20 m for the deepest 720 m continuous reflector (Jurassic age basalt flow?) to about 8 m for the reflector at about 320 m depth. The reflectors above appear gently upwarped. The fault to the west dips steeply to the east and displays small (< 10 m) displacements to within about 340 m of the surface. These two faults are nearly centered on the upwarped Upper Cretaceous sediments along the ZRA, which suggests that the two faults on the EXXON profile are part of a buried fault system, uplift along which produced the upwarped Upper Cretaceous sediments and the ZRA.

A seismic reflection profile was acquired across the ZRA just north of the Santee River in early 1996 by the University of Kansas [11]. This ~15 km long, high-resolution seismic reflection profile revealed two steeply dipping (both east side up) faults that are ~6.5 km apart. The fault to the west, which is nearly centered on the ZRA, displays greater vertical offsets than the fault to the east. Displacements range from ~60 m at ~770 m depth to ~15 m at ~280 m depth. The reflected events on the east side of this fault between ~260 m and 500 m depths are gently upwarped.

Three seismic reflection surveys acquired by the U.S. Geological Survey in the early 1980s near Summerville (e.g., SC-4, SC-6 and SC-10; Fig. 3) revealed three possible faults (Gants and Cooke faults and the edge of the missing "J" of Fig. 3; [6] that are nearly centered on the ZRA. The Gants and Cooke faults, both of which coincide with the linear aeromagnetic anomaly, are characterized by west-side-up offsets of about 50 m in a Jurassic age basalt layer at a depth of about 700 to 750 m [6]. Marple and Talwani [10] reinterpreted the edge of the missing "J" as an offset in the Jurassic age basalt at about 750 m depth. Three shallow, high resolution seismic reflection profiles that were acquired across the ZRA near Summerville in 1993 also revealed buried faults with small west-side-up offsets and/or upwarped sediments along the ZRA [9] (Fig. 3). A few of these coincide with the linear magnetic anomaly (Fig. 3), which suggests the presence of a NNE-trending buried fault zone beneath the ZRA near Summerville.

An integration of the various data presented above leads to the conclusion that there is a broad ~200 km long, NNE trending fault zone below the Coastal Plain sediments. This feature is the southernmost segment of an ~600 km long fault zone extending northeast to the Virginia Coastal Plain, named the East Coast Fault System (ECFS) [11]. In the next section we will examine various data that suggest neotectonic activity on the South Carolina segment of the ECFS.

EVIDENCE OF NEOTECTONIC ACTIVITY

Evidence of neotectonic activity is divided into loosely defined time scales, which cover the last 1,000,000 years. These different lines of evidence include the ZRA, upwarped Plio-Pleistocene deposits, paleoearthquakes, releveling data, current seismicity and GPS investigations.

River Morphology - Methodology
River morphology was used to search for buried active faults in the Coastal Plain because alluvial rivers often produce certain local morphological changes in response to local tectonic uplift or subsidence [18] (Table 1). Here we discuss the general effect of uplift on Coastal Plain rivers because the ENE oriented compressive maximum horizontal stress field in eastern US favors uplift rather than subsidence, along buried active faults in the Atlantic Coastal Plain. In areas of uplift, steepening of the valley floor slope immediately downstream from the uplift axis results in an increase in stream power (the product of stream discharge and channel slope) [18]. The initial response of the meandering river is to maintain its equilibrium slope along the steepened reach by increasing its sinuosity. As the sinuosity

increases, the sediment carrying capacity of the stream decreases and aggradation occurs. The increase in stream power also results in the development of local knickpoints and channel entrenchment that progress upstream. Upstream from the uplift axis, the decreased valley slope and corresponding loss of stream power results in a decrease in the sediment carrying capacity of the stream, which subsequently causes channel aggradation. Continued aggradation of the channel reduces the channel capacity, which may induce flooding and the development of anastomosing patterns upstream from the uplift axis. Continued uplift through time results in upwarped floodplains and terraces across the central area of uplift.

Table 1: Response of experimental river meandering channels to uplift and subsidence. (Modified from [18])

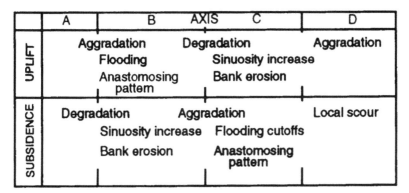

	A	B	AXIS C	D
UPLIFT	Aggradation Flooding Anastomosing pattern	Degradation Sinuosity increase Bank erosion		Aggradation
SUBSIDENCE	Degradation	Aggradation Sinuosity increase Bank erosion	Flooding cutoffs Anastomosing pattern	Local scour

The size of a river channel, based on discharge or energy, also affects the type and degree of its response to tectonic uplift [18]. In general, small channels are unable to incise across the uplift axis. Rivers of intermediate size may incise across the axis, but their longitudinal profiles still exhibit upward-displaced convexities. The large rivers traversing an area may keep pace with the uplift by incision and their projected channel profiles are relatively straight, although their terraces and valley floors are upwarped. Their floodplains do not display anastomosing patterns above the uplift axis. Below the uplift axis the larger rivers develop new floodplains as their former floodplains become low terraces. In areas where uplift is sufficiently rapid, some smaller rivers become diverted along new courses away from the area of uplift [18].

Data Sources: U.S. Geological Survey 7.5 minute and 1:100,000 scale topographic maps, color infrared aerial photographs and Landsat imagery were used to investigate the rivers traversing the study area. The Landsat imagery and aerial photos were used to search for anomalous drainage patterns and lineaments whereas the 7.5 minute topographic maps were used to produce terrain profiles and cross- valley and longitudinal valley profiles for the rivers. Most topographic maps used to construct the various profiles had contour intervals of 5 feet. In certain areas, however, the topographic maps had larger contour intervals, which would have precluded detection of the more subtle river anomalies.

Longitudinal profiles of rivers and fluvial terraces were used to infer tectonic deformation [e.g., 3, 18]. The profiles were constructed along the middle of the floodplain or terrace and then projected to the middle of the valley.

The local uplift rate can be estimated from longitudinal profiles by comparing the amounts of vertical deformation of fluvial terraces and floodplains to their ages of formation [2, 18]. These ages were taken from surficial geologic maps where available. Assuming the original longitudinal profile before uplift is generally smooth and concave upward, the maximum amount of uplift can be estimated. Cross-valley profiles   were constructed transverse to the trends of river valleys and were used to identify areas of increased channel incision and anomalous changes in cross-valley shape.

Calculation of Channel Sinuosity: Any tectonic deformation that changes the slope of a river valley results in a corresponding change in channel sinuosity (ratio of channel length to valley length along the valley axis) to maintain its equilibrium channel slope [e.g., 3]. In this study, the valley length was broken into segments of equal length and a value of sinuosity was calculated for each segment. The segment length was at least twice the meander wavelength of the channel. No sinusitis were calculated for the portions of rivers characterized by anastomosing patterns because no main channel could be identified along this part of the river valley.

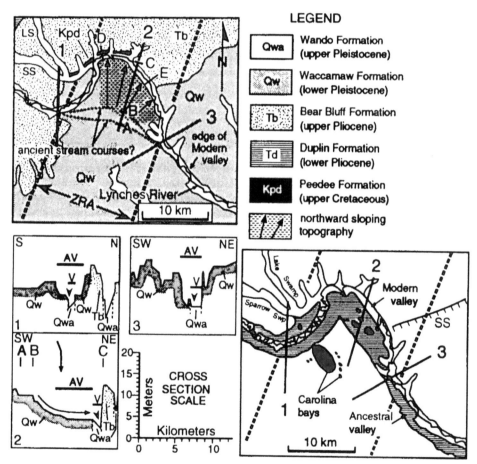

Fig. 4   Topographic profiles across the modern (V) and ancestral (AV) Lynches River valleys (denoted by thick horizontal lines in cross sections) with surficial geology taken from Owens (1989) (upper left map and cross sections). Arrowheads indicate channel location. ZRA(S) between parallel dashed lines. Note on profile 2 (lower left cross section) the NNE-downward tilt of the terrain south of the Lynches River (denoted by arrows and shaded area in location map). The reach of the modern valley of the Lynches River where the channel is ingrown against the north side of its valley is located between locations D and E in upper left map. Profiles do not include channel depth beneath water surface.   Depth of sedimentary units in the cross-sections are not the actual depths of the sediments.

River Morphology - Observations

The Lynches River valley displays the greatest variety of river anomalies associated with the ZRA (Fig. 2). For example, the river is ingrown against the northern side of the valley curve for ~8 km. The sinuosity along this reach decreases significantly for ~9 km, then increases farther downstream [10]. The longitudinal profile of its floodplain displays a convexity that is displaced upward ~3 m (Fig. 3 [10]). Its cross-valley shape also changes where it traverses the ZRA. In contrast to the generally symmetric cross-valley shape of the Lynches River valley upstream and downstream from the ZRA, its ancestral and modern valley floors along and just south of the curve in its modern valley display a NNE-downward cross-valley tilt (Fig. 4). The direction of this tilt is opposite the regional southwest downward tilt associated with the southern flank of the Cape Fear arch. The topographically distinct ancestral valley, which the modern valley is entrenched ~2-3 m, also displays a NNE convex, arc-shaped curve where it traverses the ZRA (Fig. 4).

The Santa River's floodplain north of Summerville is upwarped ~2 m across the ZRA (Fig. 3 in [10]). Along the downstream side of the valley curve and the upwarped area, ~10 km of the floodplain is heavily dissected (Fig. 2).

The Black River's floodplain, located farther north, is heavily dissected along the upstream side of its valley curve (Fig. 2). Upstream from this area, the Black River's floodplain is characterized by anastomosing patterns whereas along and downstream from the curve it is characterized by a meandering channel (Fig. 2).

The tributaries that drain into the Black River valley along the north side of the valley curve also display anastomosing patterns.

The Pee Dee River valley located east of Florence undergoes several abrupt changes where it traverses the northern end of the ZRA. Downstream from the ZRA, the river valley exhibits a noticeable southwest downward cross-valley tilt with well developed, unpaired terraces on its northeast side that decrease in age and elevation toward the southwest [19]. In contrast, the valley upstream from the ZRA is nearly symmetric with fewer, relatively narrow terraces on both sides of the valley. The youngest terrace into which the Pee Dee River is slightly entrenched, narrows from ~7 to 10 km upstream from the ZRA to ~1.5 to 5 km along and downstream from the ZRA. The longitudinal profile of this terrace displays a convexity that is displaced upward ~3 m where it traverses the ZRA (Fig. 3 in[10]). The river's meandering pattern also changes abruptly where it traverses the ZRA. The river upstream from the ZRA currently meanders across the center of the valley floor whereas downstream from the ZRA it is located mostly against the southwestern valley wall. Where it traverses the ZRA, the channel becomes nearly straight [10] and is located against the northern side of the Wando terrace.

Topography Along the ZRA

Investigation of the topography along the ZRA revealed an ~30 km long, 13 to 20 km wide, topographic high along the ZRA between Lake Moultrie and the Ashley River that is -20 m higher than the surrounding Coastal Plain (Fig. 3). Smaller, more localized topographic highs of about 5 m were also noted south of the Ashley River along the ZRA (Fig. 3) [10]. No obvious topographic highs were observed along the ZRA north of Lake Moultrie.

The topography along the axis of the ZRA generally increases in elevation toward the north-

northeast. It is cut by a number of fluvial valleys forming topographic low areas, such as that presently occupied by Lake Moultrie (Fig. 5). The topographically low area beneath Lake Moultrie was cut by the drainage network associated with a tributary of the Cooper River, Biggin Swamp, which drained the area beneath Lake Moultrie prior to its damming in 1944. The NNE convex, arc- shaped curves in the Santee, Black and Lynches river valleys are deflected along the ZRA toward the direction of generally higher topography and the main area of uplift associated with the Cape Fear arch (Fig. 5). This observation, combined with the NNE downward tilt of the terrain on the south side of the modern valley of the Lynches River, suggests a gentle RTNE downward tilting of their valley floors along the ZRA.

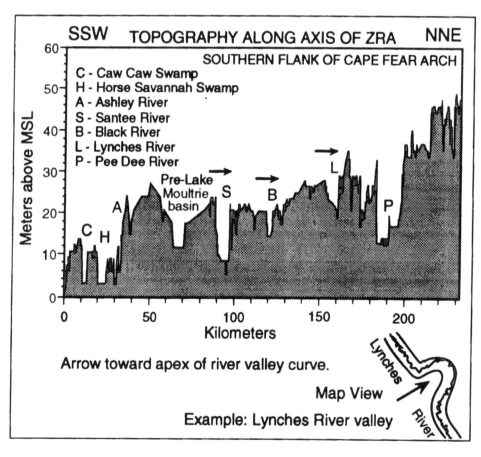

Fig. 5 Topographic profile approximately along the axis of the ZRA. Note that the NNE convex curves in the river valleys along the ZRA(S) are deflected toward generally increasing topography and uplift from the Cape Fear arch

The minimum age of tectonic activity can be inferred from the ages and patterns of surficial sediments along the arc-shaped curves in the river valleys where they traverse the ZRA. Carolina Bays formed on the ancestral valley along the ZRA between 100,000 and 200,000 years before present. Based on their presence, the NNE downward cross-valley tilt along the ancestral valley of the Lynches River where it traverses the ZRA is at least 200 ka (10^3 years) old [11]. This observation suggests that the minimum age of buried faulting associated with this tilt is at least 200 ka [11].

Shallow Stratigraphy
The NNE-trending early Pleistocene Summerville barrier spit deposit lies between Lake Moultrie and Summerville. It is 2-3 m higher than the surrounding area. The elevated barrier and the underlying shallow marine sediments were deposited on a NNE-trending buried structural high in the pre-Pliocene-Pleistocene surface [23]. A clay unit within the Black Creek formation is upwarped along the ZRA between the Lynches and Santee rivers. This NNE-trending area of upwarped, Upper Cretaceous sediments is collinear with the pre-Pliocene-Pleistocene high beneath the Summerville barrier. These observations suggest a Pleistocene age for tectonic activity on the Summerville barrier.

Paleoseismicity
The 1886 Charleston earthquake was associated with widespread sandblows, which were caused by the liquefaction of buried sands. Sandblows provide an indirect evidence of prehistoric earthquakes. By dating trapped organic sediments within the sandblows and those that are cut by them it is possible to estimate the date of occurrence of the prehistoric earthquakes.

Since 1983 over 150 sandblows have been discovered in the Coastal Plain of South Carolina and more than two-thirds of those have been age-dated. Paleoliquefaction activity is restricted by the depth to the water table. Various data suggest that prior to 5000 years before present, the water table was too low to produce sandblows that could reach the surface.

Analyses of all available paleoseismological data suggest that there may have been at least six and possibly seven paleoearthquakes in the outer South Carolina Coastal Plain. The only paleoliquefaction features that have been found in South Carolina lie along the coast northeast and southwest of Charleston [1, 16, 24]. The ages of the paleoearthquakes are 110, 546 ± 17, 1001 ± 33, 1641 ± 89, 3548 ± 66, 5038 ± 166 and 5300 to 6300 years before present [22]. The discovery of sandblows of similar ages near Charleston and to its northeast and southwest argue for a source near Charleston. However, sandblows for the event dated 1641 ± 89 were found only in the north near Georgetown and Myrtle Beach but not near Charleston, which argues for a seismic source north of Charleston. These observations suggest that the seismic source associated with the seismicity near Charleston extends to the northeast, possibly along the ECFS. No evidence of prehistoric earthquakes was found towards the northwest.

CONCLUSIONS

Based on an integration of a variety of data we suggest the presence of a buried ~200 km long fault zone beneath the Coastal Plain of South Carolina. The evidence of tectonic activity on this feature from Holocene time to 1,000,000 years before present comes from a variety of observations. The upwarped floodplains along the Santee, Lynches and Pee Dee rivers (Fig. 1) indicate tectonic activity between about 100,000 years before present and Holocene time. Observations of surficial deposits, combined with changes in the cross-valley shapes of the Santee and Lynches river valleys along their arc-shaped curves suggest local uplift along the ZRA since at least Penholoway time (~750,000 years, [12]) and continuing through Holocene time.

Paleoliquefaction data suggest that there was earthquake activity at least as far back as 5,000 years. Historical seismicity has been documented for about the last 300 years. In view of the current seismicity we conclude that there has been tectonic activity for at least the last 5,000 years. Additional evidence of local tectonic activity for the last 100 years comes from an evaluation of the releveling data in the area [e.g., 15].

In the sections above we showed that in the absence of fault scarps and other surficial manifestations of large faults it is still possible to delineate long buried faults in alluvial settings of intraplate regions. This was illustrated with data from the Coastal Plain of South Carolina, the location of the destructive 1886 Charleston, South Carolina, earthquake. Integration of a variety of data support the existence of a NNE-trending buried fault zone along the ZRA. The length of this NNE-trending fault zone is adequate to generate the M_w 7.3 estimated for the 1886 Charleston earthquake. A variety of data indicate that there has been tectonic activity on this fault zone for at least 1,000,000 years. Currently, the most seismically active part of this fault zone is its southern end where it intersects with the Ashley River fault. This intersection provides a location for stress accumulation due to plate tectonic forces and the pursuant seismicity.

REFERENCES

1. D.C. Amick and R. Gelinas, The search for evidence of large prehistoric earthquakes along the Atlantic seaboard: *Science* 251, no. 4994, 655-658 (1991).
2. A.W. Burnett, Alluvial stream response to neotectonics in the lower Mississippi valley [M.S. thesis]: Fort Collins, Colorado, *Colorado State University*, 182 p. (1982).
3. A.W. Burnett and S.A. Schumm, Active tectonics and river response in Louisiana and Mississippi: *Science* 222, 49-50 (1983).
4. C.E. Dutton, The Charleston earthquake of August 31, 1886: *U.S.G.S. Ninth Annual Report* 1887-1888, 528 p. (1889).
5. G.S. Gohn, ed., Studies related to the Charleston, South Carolina, earthquake of 1886-tectonics and seismicity: *U.S.G.S. Prof. Paper* 1313, A1-T11, 8 plates (1983).
6. R.M. Hamilton, J.C. Behrendt and H.D. Ackermann, Land multichannel seismic reflection evidence for tectonic features near Charleston, South Carolina; in G.S. Gohn, ed., Studies related to the Charleston, South Carolina, earthquake of 1886; tectonics and seismicity: *U.S.G.S. Prof. Paper* 1313, 11-118 (1983).
7. A.C. Johnston, Seismic moment assessment of earthquakes in stable continental regions-III. New Madrid 1811-1812, Charleston 1886 and Lisbon 1755: *Geophys. Jour. Intern.* 126, 314-344 (1996).
8. S. Madabhushi and P. Talwani, Fault plane solutions and relocations of recent earthquakes in Middleton Place Summerville seismic zone near Charleston, South Carolina: *Bull. Seis. Soc. Amer.* 83, no. 5, 1442-1466 (1993).
9. R.T. Marple, Discovery of a possible seismogenic fault system beneath the Coastal Plain of South and North Carolina from an integration of river morphology and geological and geophysical data [Ph.D. dissert.]: University of South Carolina, Columbia, 354 p., 13 plates (1994).
10. R.T. Marple and P. Talwani, Evidence of possible tectonic upwarping along the South Carolina coastal plain from an examination of river morphology and elevation data: *Geology* 21, no. 7, 651-654 (1993).
11. R.T. Marple and P. Talwani, Neotectonics of a major buried fault system in the Coastal Plain of the Carolinas and Virginia: Cause of the 1886 Charleston, South Carolina earthquake? *Geol. Soc. Amer.* Bull. (in review).
12. L. McCartan, R.E. Weems and E.M. Lemon Jr., Quaternary stratigraphy in the vicinity of Charleston, South Carolina, and its relationship to local seismicity and regional tectonism, in Studies Related to the Charleston, South Carolina, Earthquake of 1886-Neogene and Quaternary Lithostratigraphy and Biostratigraphy: *U.S.G.S. Prof. Paper* 1367-A, A1-A39 (1990).
13. J.P. Owens, Geologic map of the Cape Fear region, Florence $1° \times 2°$ quadrangle and northern half of the Georgetown $1° \times 2°$ quadrangle, North Carolina and South Carolina: *U.S.G.S. Misc. Invest. Map* I- 1948-A, scale 1:250 000, 2 sheets (1989).
14. J.D. Phillips, Buried structures at the northern end of the early Mesozoic South Georgia basin, South Carolina, as interpreted from aeromagnetic data, in A.J. Froelich and G.R. Robinson, eds., Studies of the early Mesozoic basins of the eastern United States: *U.S.G.S. Bull.* 1776, 248-252 (1988).

15. C.M. Poley and P. Talwani, Recent vertical crustal movements near Charleston, South Carolina: Jour. *Geophys. Res.* **91**, 9056-9066 (1986).
16. C.P. Rajendran and P. Talwani, Paleoseismic indicators near Bluffton, South Carolina: An appraisal of their tectonic implications: *Geology* **21**, 987-990 (1993).
17. D.W. Rankin, ed., Studies related to the Charleston, South Carolina, earthquake of 1886 - A preliminary report: *U.S.G.S. Prof. Paper* 1028, 204 p., (1977).
18. S.A. Schumm, Alluvial river response to active tectonics, in Studies in Geophysics: Active Tectonics: Washington, D.C., *National Academy Press*, 80-94 (1986).
19. D.R. Soller and H.H. Mills, 1991, Surficial geology and Geomorphology, in J.W. Horton, Jr. and V.A. Zullo, eds., *The Geology of the Carolinas*: Carolina Geological Society Fiftieth Anniversary Volume: Knoxville, The University of Tennessee Press, 290-308 (1991).
20. P. Talwani, An internally consistent pattern of seismicity near Charleston, South Carolina: *Geology* **10**, 655-658 (1982).
21. P. Talwani, Seismotectonics of the Charleston region, in *Proceedings, National Conference on Earthquake Engineering, 3rd*: Earthquake Engineering Research Institute **1**, 15-24 (1986).
22. P. Talwani and D.C. Amick [in preparation].
23. R.E. Weems, and S.F. Obermeier, The 1886 Charleston earthquake --An overview of geologic studies, *Proceedings, Water Reactor Safety Information Meeting,17th*: U.S. Nuclear Regulatory Commission NUREG/CP-0105, 289-313 (1990).
24. R.E. Weems, S.F. Obermeier, J.J. Pavich, G.S. Gohn, M. Rubin, R.L. Phipps and R.B. Jacobson, 1986, Evidence for three moderate to large prehistoric Holocene earthquakes near Charleston, South Carolina: *Proceedings, 3rd U.S. National Conference on Earthquake Engineering*, Earthquake Engineering Research Institute **1**, 197-208 (1986).

Proc. 30* Int'l. Geol. Congr., Vol. 5 pp. 63-71
Ye Hong (Ed)
© VSP 1997

Adriatic Slab Versus Tyrrhenian Lithosphere (in the Mediterranean Puzzle): New Suggestions for an Old Problem

M. BERNABINI[1], D. DI BUCCI[2], L. ORLANDO[1], M. PAROTTO[2] AND M. TOZZI[3]

1)Dipartimento di Idraulica, Trasporti e Strade, Università degli Studi di Roma "La Sapienza", Rome, Italy.
2)Dipartimento di Scienze Geologiche,III Università degli Studi di Roma, Rome, Italy.
3)CNR - C. S. per il Quaternario e l'Evoluzione ambientale, Rome, Italy.

Abstract

Available geological and geophysical data do not yet provide a clear picture of the deep crustal structure beneath Central Italy. The National Research Project CROP, and specifically the CROP 11 sub-Project will focus on the crustal structure of Central Italy.

Gravimetric studies have already been performed. By a new interpretation of already exist data, they have allowed the development of a two-dimensional model of the crustal and upper mantle setting, along a cross section from Tyrrhenian to Adriatic Sea (corresponding to the acquisition trace of the CROP 11 seismic line).

This model suggests the existence of a Tyrrhenian lithospheric sector having lower densities with respect to the Adriatic sector. These two sectors are separated by an intermediate zone, where a lighter wedge of the upper mantle has been hypothesized. It has been interpreted as part of a slab of Adriatic crust underneath the Tyrrhenian Moho.

On this basis, for the comprehension of the regional deep setting, a 3D extension of the 2D previous model is now necessary. This will provide better evaluation of the lateral development of the identified structures. The first step of this process is the subject of this report. It consists in the study of the areal development of the residual gravimetric anomalies, exclusively due to deep structures. We started from the Bouguer anomaly map (2.67 g/cm^3) and removed the effects of superficial structures applying the stripping-off method to the topographic surface.

In this way, a new residual map for the Central Italy has been obtained. A prominent N-S direction can be recognised in the northern-central part of the study area, related to the flexed part of the Adriatic slab. It characterises the crustal portion and confirms that the CROP 11 deep seismic line is oriented properly. A more complex deep structure seems to be localized in the central-southern part of the region investigated, suggesting an important change going from north to south in the lower crust and upper mantle (these data also agree with the DSS and tomography data). Finally, the residual map shows that isolines trend W-E in the middle of the study area, in both Tyrrhenian and Adriatic side, as well as in the core of the Apenninic chain. It is difficult, at the moment, to find a real geological correspondence for these sets of isolines including the existence of important W-E structures, as the 41° parallel or the meso-adriatic belt, and they constitute an interesting theme for debate.

Keywords: Adriatic Slab, Tyrrhenian lithosphere, gravimetric study

INTRODUCTION

Most of the mountain chains present in the Mediterranean area are segmented in more than one arcuate fold and thrust belt [9,10,17,19]. This is particularly evident for the Apenninic chain, where there are two major arcs, with different direction and rate of shortening [4,7]. This superficial setting reflects different characters of the slab in the Northern Apennines with respect to the Southern Apennines, as shown by geophysical data (seismic tomography, DSS, deep seismic lines, etc.; [1,20]). In this general framework, the deep geological setting at the junction between the northern and southern Apenninic arcs is still unknown.

To increase the knowledge about the subsurface, major Italian research authorities (CNR, Universities, ENEL, AGIP) are co-operating in the deep seismic project CROP (CROsta Profonda, deep crust), similar to those of other countries, as BIRPS, COCORP, DEKORP, ECORS, NFP20. The CROP 11 deep seismic line, from the Tyrrhenian to the Adriatic Sea, is the central one (Fig. 1, after[6] and[3]), and investigates that portion of Apenninic crust that lays between the northern and southern arcs.

The preliminary studies for this line have represented a good chance to review all existing geological and geophysical data [6,22]. In this context, the gravimetric approach plays an important role, because it reflects the distribution of crustal volumes in depth. A two-dimensional gravimetric model from Tyrrhenian to Adriatic Sea, corresponding to the CROP 11 section, has already been reported, pointing out significant differences of density between Tyrrhenian and Adriatic lithosphere and suggesting the presence of a wedge of low density in the central part of the upper mantle[3].

Starting from these results, the aim of this work is to evaluate the three-dimensional development of the deep structures identified by the 2D modelling, giving further constraints for the best orientation of the CROP 11 deep seismic line before and during its acquisition.

In order to do this, residual gravimetric anomalies, exclusively due to deep structures, have been studied. We started from the Bouguer anomaly map $(2.67g/cm^3)$ and removed the effects of superficial structures by the stripping-off method [2,3].

This resulted in a residual map for the Central Italy. An important N-S direction can be recognized in the northern-central part of the study area, related to the flexed part of the slab, and confirming that the deep seismic line is oriented properly. A more complex deep setting seems to be localized in the central-southern part of the region investigated, suggesting an important change going from north to south in the lower crust and upper mantle.

GEODYNAMIC AND GEOLOGICAL FRAMEWORK

The Italian peninsula is characterised by two main geological domains. The first, represented by the Apenninic chain system, is an orogenic belt developing since the Late Cretaceous as a result of the Europe-Africa collision and the passive sinking of the Adriatic-Ionian lithosphere. The Apenninic chain is formed by two main arcs: the northern one and the southern one, verging respectively toward NE and SE with different rates of shortening [4,7,19]. The second domain corresponds to the Tyrrhenian margin. Here, along the western margin of the Italian peninsula, superposed on the compressive structures, extensional features related to the opening of the Tyrrhenian Sea have developed since Upper Miocene. These two crustal domains are different from a geological and a geophysical point of view (summarized, among the others, in Cavinato et alii[6].

In both Northern and Southern Apenninic arcs, these crustal domains are separated in correspondence of slabs, identified by means of geophysical studies (earthquake hypocentres, seismic tomography, results of other CROP seismic lines, etc. ; [1, 6, 22]), while available

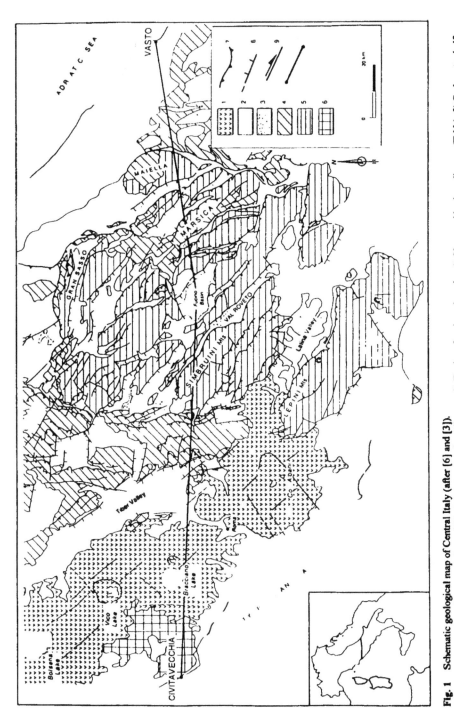

Fig. 1 Schematic geological map of Central Italy (after [6] and [3]).
1) Volcanics (Q). 2) Marine and continental N_2-Q_2 post-orogenic sediments. 3) Tertiary foredeep deposits. 4) Slope and basin sediments (T-N_1). 5) Carbonate shelf and shelf-edge sediments (T-N_1). 6) Tuscan nappe. 7) Thrust faults. 8) Normal faults. 9) Strike-slip faults. 10) Trace of CROP 11 deep seismic line.

geological and geophysical data do not yet provide a clear picture of the deep crustal structure beneath Central Italy. For this area, silent from a seismological point of view, the presence of a slab has been inferred only by matching gravimetric and DSS (deep seismic sounding) data (Fig. 2 [3;5;20]).

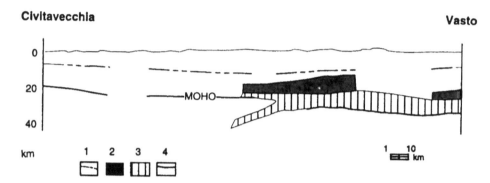

Fig. 2   Simplified 2D interpretation of an Adriatic slab (Vasto) underneath the Tyrrhenian Moho (Civitavecchia) in Central Italy. 1) Aeromagnetic basement; 2) low velocity layer in the lower crust; 3) higher velocity layer in the lower crust; 4) Moho (after [3], modified and simplified).

An important test for the two-dimensional model obtained from these data is the recognition of the continuity of the identified structures on both sides of the cross-section. In order to do this, the first step is shown in this paper. It consists of the study of the areal development of gravimetric anomalies exclusively due to deep structures. Matching the trends visible on the residual map with the results of the two-dimensional model already existent, a three-dimensional approach is here attempted.

METHODOLOGY

Geological aspects
The surface geology of Central Italy is characterized by the presence of four main provinces, essentially part of the Apennines orogenic system and defined by different stratigraphic sequences and tectonic setting. They are, from the north-east to the south-west: 1) a deformed intraorogenic foreland (Apulia-Adriatic); 2) a deformed foredeep (Adriatic trough); 3) a thrust belt (Central Apennines chain); 4) a hinterland in extension (Tyrrhenian basin).

The correct evaluation of rock distribution and thickness in Central Italy has represented the first step to eliminate the effects of the superficial geology from the Bouguer anomalies.

1) The Apulia-Adriatic foreland is constituted by basin and platform carbonate sediments, from the Triassic to the Miocene in age, here outcropping in the Gargano area.

2) The Adriatic trough is filled by synorogenic siliciclastic deposits, some thousand metres thick, whose deposition started in Messinian times.

3) The Central Apennines chain is a pile of thrust sheets that developed in Miocene and, at least, Pliocene times toward the Apulia-Adriatic foreland. Different paleogeographic units,

developed along the southern margin of Thethys Ocean in meso-cenozoic times, are deformed in these thrust sheets. They are (Fig. 1): a) the Apulia carbonate platform, outcropping, for example, in the Maiella Mt.; b) the Genzana basin, in the eastern Marsica, characterized by calcareousand marly-calcareous deposits. Toward SE, these deposits become clay-rich, defining the passage to the Molise basin (recognizable to the SE of the Maiella Mt.); c) the Latium-Abruzzo carbonate platform, outcropping in the Lepini and Simbruini Mts. Moreover, along the major valleys of this portion of the chain, synorogenic siliciclastic deposits, part of Upper Tortonian-Messinian foredeeps, have been preserved.

The core of the Apenninic thrust belt is also overprinted by successive strike-slip and/or extensional tectonics. The extension, in particular, is responsible for the formation of large intramontane basins (e. g. Fucino basin; [6]) filled by some hundred metres of fluvial and/or lacustrine sediments.

4) The Tyrrhenian province is strongly affected by extension tectonics, due to the opening of the Tyrrhenian Sea. This extension is responsible for the presence of huge volcanic complexes (Vico, Bracciano, Colli Albani), whose deposits cover the clayey and sandy deposits of the Plio-Quaternary marine transgression. Moreover, at the western end of the trace of the CROP 11 seismic line, siliciclastic arenaceous-clayey deposits crop out widely. They are part of a more internal paleogeographic domain, well developed in the Northern Apennines (Tuscan nappe), and can be found also in deep wells drilled off-shore, near the coast [4].

The Bouguer anomaly data used in this work have been calculated for a density of 2.67g/cm^3. This density is usually accepted for outcropping carbonate deposits, and has been chosen because this lithology is widely exposed in the study area. This means that the superficial carbonate units do not affect the anomaly trends.

On the other hand, this is not true for all the other deposits, generally with lower densities. Due to the presence of these other deposits, the regional trend of the Bouguer anomaly, originated by deep structures, is overprinted by the local effects of the surface geology.

In order to eliminate them, the second step was the reconstruction of the volumes constituted by such deposits. In this part of the work, the help of F. Cifelli and F. Funiciello was essential. Mainly based on published data (especially maps, geological cross-sections and deep wells), isobaths and isopachs maps, as well as detailed cross-sections, have been made for the most important areas occupied by deposits with densities different from 2.67 g/cm^3. These areas are: the Tyrrhenian margin (Tuscan nappe, Plio-Quaternary marine sediments, volcanic deposits), the Latina valley (siliciclastic deposits), the Fucino plain (lacustrine and siliciclastic deposits), the Molise thrust sheets (clayey basinal deposits) and the Adriatic trough (siliciclastic deposits). Several parallel cross-sections, generally E-W striking, have been prepared at 1:250,000 scale.

The third step has consisted in defining each section by a number of UTM co-ordinates (and relative elevations), realizing the data base used for the successive stripping-off [2].

Geophysical aspects
Bouguer anomaly values calculated for a constant density of 2.67 g/cm^3 from the National Gravimetric Data Base, distributed on a grid of 3 km × 3 km, have been used for the present work. This grid was produced by interpolating data from all gravimetric stations in Italy (density of distribution of about 1 station per km^2).

The study area lies between 42°40' N and 41°20' N of latitude. The Bouguer anomaly map for this area is presented in Fig. 3. From a regional point of view, in its northern portion, the map shows a negative gradient from the Tyrrhenian side to the Adriatic, with a minimum value corresponding to maximum thickness of the Adriatic foredeep. In the southern portion, a minimum corresponding to the external part of the chain can be observed. This minimum arises with a positive gradient toward the Apulia foreland. Local points of minimum are also due to the presence of intramontane basins.

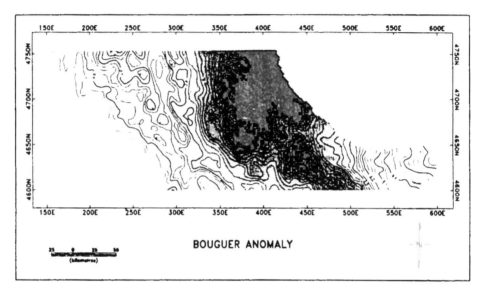

BOUGUER ANOMALY

Fig. 3   Bouguer anomaly map for the study area (density: 2.67 g/cm³).

As the Bouguer anomalies result from the effects of both superficial and deep structures, in order to obtain the gravimetric anomalies related exclusively to the deep ones, the effects of superficial features have been filtered. Practically, for each volume of rock identified by the described sets of geological cross-sections, an appropriate constant density has been attributed. The densities used are the following: 2.5 g/cm³ for the siliceous-marly formations (Molise units); 2.4 g/cm³ for the siliciclastic deposits (flysch); 2.3 g/cm³ for the clayey marine Plio-Quaternary sediments; 2.1 g/cm³ for the intramontane basin fluvial-lacustrine deposits; 2.0 g/cm³ for the volcanic formations. A density of 2.67 g/cm³ has been attributed to the Meso-Cenozoic carbonate sediments, and for this reason they have not been considered in the stripping-off. The gravimetric contribution of every volume of rock to the Bouguer anomaly map has been calculated and removed by using a 3D gravimetric software [14].

RESULTS

The residual map obtained in this way is shown in Fig. 4. From a general point of view, we can observe that the effects of the Adriatic foredeep and of the Fucino intramontane basin are disappeared. Residual positive anomalies can be still observed all along the Tyrrhenian margin and in correspondence with the Gargano promontory. The core of the chain is, instead, characterized by a negative residual anomaly, that is particularly evident in the northern-central part of the map (23-25 mGal), where it geographically corresponds to the Gran Sasso massif. Some intramontane basins also occur in this region, but were not modelled as the thickness and density of their sediments could be responsible only for a minor part of this anomaly (3-5 mGal). Therefore, the observed negative residual anomaly is mainly due to deep structures, that could be interpreted as shown in the existing two-dimensional model[3], i. e. by the presence of a wedge of low density in the upper mantle.

Going from the minimum to the positive anomalies of the Tyrrhenian region, the isolines show a well defined N-S trend, that suggests a similar direction for the related deep structures, i. e. the underthrusting Adriatic slab.

Toward the south, the negative anomaly strongly decreases while the isolines assume an irregular pattern. This seems to sign an important variation in the deep structures, because it becomes impossible to follow toward the south the structures identified in the northern-central part of the study area.

Finally, the Tyrrhenian margin appears divided in two sectors by a W-E trend of the isolines, that show a positive gradient toward the south. A direct correlation with geological features is difficult. The region from the Tuscany to the Gulf of Naples is, in fact, characterized by high values of heat flow and temperature gradient [15], by the presence of igneous rocks and important volcanic complexes, and in general by extensional geological features associated with the spreading of the Tyrrhenian Sea. These conditions can modify the properties of the deep rocks causing density variations that must be still evaluated in detail.

DISCUSSION AND GEODYNAMIC IMPLICATIONS

Matching the residual map presented in this work with the two-dimensional gravimetric model published in Bernabini et alii [3] (Figs 2 and 4), an initial three-dimensional interpretation of the Central Italy lithospheric features can be attempted.

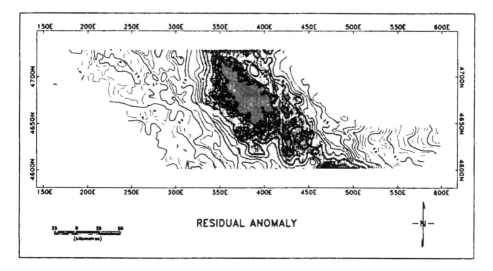

Fig. 4   Residual map of gravimetric anomalies due to deep crust and upper mantle structures.

First of all, in Bernabini et alii [3] the two-dimensional gravimetric modelling allowed an interpretation of the low density wedge below the Moho as a low density flake of Adriatic lithosphere subducting towards the Tyrrhenian Sea. The residual anomalies presented here suggest that this structure continues with a N-S trending for about 100 km. To the north it diminishes, while to the south it seems to be abruptly interrupted, distinguishing the examined portion of the Apenninic lithosphere with respect to the northern and the southern ones, and supporting as appropriate the W-E direction chosen for the CROP 11 deep seismic line. For a long time, the presence and direction of an Adriatic slab underneath Central Italy has constituted a theme open to debate. In the light of our results, such presence seems to be confirmed. On the other hand, due to lack of deep earthquakes in Central Italy, it is unsupported by any seismological data, while evidence from earthquake hypocentres makes the interpretation of the presence of a slab subducting toward the west for Northern and Southern Apennines acceptable [1,13]. Geometric and geodynamic difficulties arise in interpreting the gravimetric trends of the region corresponding to the deep Adriatic thrust hypothesized underneath Central Italy. The abrupt change of trend from N-S to NW-SE could possibly be interpreted as a ductile change of direction in the underthrusting slab. This could be explained in the context of a ductile deformation of the entire Adria plate underthrusting margin, all along the Italian peninsula (Northern Apennines and Calabrian Arc). On the other hand, this change of direction could be caused by a lithospheric tear fault, cutting off the original continuity of the Adria plate margin (similar tears have been hypothesized by Patacca et alii, [19]. In regards to this last possibility, Favali et alii [11] and Gambini & Tozzi [12] recently pointed out the regional significance of important W-E trending transcurrent faults affecting the Adriatic foreland. These faults are interpreted as due to the different geodynamic behaviour of the southern part of the Adria plate, thicker and more rigid with respect to the northern one. Satellite image and shadow lineaments analyses, as well as seismicity and drainage network, also suggest the presence of important E-W trending lineament domains, that trend across the Adriatic Sea toward the inner Apenninic chain [8,11,23]. Moreover, abrupt lithospheric breaks seem to be necessary to explain the different age and facies distribution of Central-Northern Italy foredeeps [7]. Finally, similar lithospheric tears have also been hypothesized in other geodynamic reconstructions. The Moho depth maps [16,18,20] show a very important discontinuity in the Adriatic Moho flexural trend underneath Central Italy. More complete geodynamic reconstructions of this sector of the Italian peninsula, also based, for example, on DSS data, are not available, due to the present lack of data sets.

Secondly, the N-S setting can be recognized only in the central-northern part of the study area (Fig. 4), being abruptly cut southward, where it is almost impossible to identify any well characterized structure. This clear difference suggests a developed deep separation with respect to the southern part of Italian peninsula, just where surface data [19] indicate the junction between the two major arcs of the Apennines. This also corresponds to a very complex subsurface geometry that makes the gravimetric interpretation much more uncertain.

Finally, the residual anomalies show a significant W-E isoorientation in the central part of the study area. This is particularly evident along the Tyrrhenian margin and the northern sector of the Gargano promontory, but detectable in the core of the chain as well.

Even though it is difficult, at the moment, to relate these sets of isolines to known geological structures, there are some antimeridian structures reported in literature. These include the aforementioned meso-adriatic belt or the 41° parallel, near the study area [11,12,19,21].

A more exhaustive picture and more precise interpretations will come by the realization of a three-dimensional model, on which the work continues.

Acknowledgments

We thanks Claudio Cesi and Renato Ventura for the gravimetric data base and for their help; Francesca Cifelli and Francesca Funiciello for their collaboration in the realization of the residual map; Chuck Cluth for strongly improving the English of the manuscript.

REFERENCES

1. A. Amato and G. Selvaggi - Terremoti crostali e sub-crostali nell'Appennino settentrionale. In: *Studi Geologici Camerti, volume speciale* (1991/1), CROP 03, 75-82. (1991)
2. M. Bernabini, P. Favaro and L. Orlando- Density anomaly and its consequence. *J. Applied Geophysics*, 32, 184-197. (1994)
3. M. Bernabini, D. Di Bucci, L. Orlando, M. Parotto and M. Tozzi -Gravimetric evidence of the deep structure in mountain building: subducted Adriatic crust beneath the Tyrrhenian Moho in Central Italy. *J. Geodynamics*, 21 (3), 223-234. (1996)
4. G. Bigi, D. Cosentino, M. Parotto, R. Sartori and P. Scandone -Structural Model of Italy. Scale 1:500,000, CNR, P. F. Geodinamica, Quaderni de "La Ricerca Scientifica", 114, 3. (1992)
5. R. Cassinis, G. Pialli, M. Broggi and M. Prosperi- Dati gravimetrici a grande scala lungo la fascia del profilo: interrogativi sull'assetto della crosta e del mantello. In: *Studi Geologici Camerti, volume speciale* (1991/1), CROP 03, 41-47. (1992)
6. G. P. Cavinato, D. Cosentino, R. Funiciello, M. Parotto, F. Salvini and M. Tozzi - Constraints and new problems for a geodynamical modelling of Central Italy (CROP 11 Civitavecchia-Vasto deep seismic line). *Boll. Geof. Teor. Appl.*, 36 (141-144), 159-174. (1994)
7. P. Cipollari and D. Cosentino - Miocene tectono-sedimentary events and geodynamic evolution of the Central Apennine (Italy). RCMNS Interim Colloquium *"Neogene Basin Evolution and Tectonics of the Mediterranean Area"*, Rabat, Morocco, Abstracts, 28-29. (1994)
8. M. Del Monte, D. Di Bucci & A. Trigari - Assetto morfotettonico della regione compresa tra la Maiella e il Mare Adriatico (Appennino abruzzese). Atti del 77° Congresso Nazionale della Societa Geologica Italiana. *Mem. Soc. Geol. It.*, 51 (in press).
9. C. Doglioni - A proposal for the kinematic modelling of W-dipping subductions. Possible applications to the Tyrrhenian-Apennines system. *Terra Nova*, 5, 423-434. (1991)
10. L. Endignoux, I. Moretti & F. Roure- Forward modelling of the Southern Apennines. *Tectonics*, 8, 1095-1104. (1989)
11. P. Favali, R. Funiciello, G. Mattietti, G. Mele & F. Salvini - An active margin across the Adriatic Sea (central Mediterranean Sea). *Tectonophysics*, 219, 109-117. (1993)
12. R. Gambini & M. Tozzi - Terziary geodynamic evolution of Southern Adria Plate. *Terra Nova*, in press. (1996)
13. D. Giardini & M. Velona - La sismicita profonda del Mar Tirreno. *Mem. Soc. Geol.. It.*, 41, 1079-1086. (1992)
14. H. J. Gotze & B. Lahmeyer - Application of three-dimensional interactive modeling in gravity magnetics. *Geophysics*, 53, 1096-1108. (1988)
15. F. Mongelli, M. Puxeddu, P. Squarci, L. Taffi & G. Zito- Il flusso di calore e l'anomalia geotermica dell'area tosco-laziale: implicazioni profonde. In: *Studi Geologici Camerti, volume speciale* (1991/1) CROP 03, 399-402. (1992)
16. C. Morelli - Risultati di 31 anni (1956-1986) di DSS e di 7 anni (1986-92) di CROP in Italia. Atti Convegno CNGTS 1994, 29-45. (1995)
17. I. Moretti & L. Royden- Deflection gravity anomalies and tectonics of doubly subducted continental lithosphere: Adriatic and Ionian Sea. *Tectonics*, 4, 875-893. (1988)
18. R. Nicolich - Crustal structures from seismic studies in the frame of the European Geotraverse (southern segment) and CROP projects. In: *The Lithosphere in Italy* (Boriani et alii Eds), 41-61. (1989)
19. E. Patacca, R. Sartori & P. Scandone- Tyrrhenian Basin and Apenninic arcs: kinematic relations since Late Tortonian times. *Mem. Soc. Geol. It.*, 45, 425-451. (1990)
20. S. Scarascia, A. Lozej & R. Cassinis- Crustal structures of the Ligurian, Tyrrhgnian and Ionian Seas and adjacent onshore areas interpreted from wide-angle seismic profiles. *Boll. Geof. Teor. Appl.*, 34 (141-144), 5-19. (1994)
21. M. Tozzi - Rotazioni e faglie trascorrenti nell'avampaese apulo: una revisione. Studi Geologici Camerti, volume speciale (1991/2), CROP 11, 249. (1992)
22. M. Tozzi, G. P. Cavinato & M. Parotto Eds- Studi preliminari all'acquisizione del profilo CROP 11 Civitavecchia-Vasto. In: *Studi Geologici Camerti, volume speciale* (1991/2), CROP 11, 441 p. (1992)
23. D. U. Wise, R. Funiciello, M. Parotto & F. Salvini - Topographic lineament swarms: Clues to their origin from domain analysis of Italy. *Geol. Soc. Am. Bull.*, 96, 952-967. (1985)

Proc. 30th Int'l. Geol. Congr., Vol. 5 pp. 73-86
Ye Hong (Ed)
© VSP 1997

Active Fault Geometry and Kinematics in Greece: The Thessaloniki (M$_s$=6.5,1978) and Kozani - Grevena (M$_s$=6.6, 1995) Earthquakes - Two Case Studies

S. PAVLIDES, D. MOUNTRAKIS, N. ZOUROS AND A. CHATZIPETROS
Department of Geology and Physical Geography, Aristotle University of Thessaloniki, GR-54006, Thessaloniki, Greece

Abstract

Two seismogenic areas in northern Greece are examined as case studies, in respect to their fault geometry, kinematics and segmentation, and their fault pattern similarities and differences. The first area, the Mygdonia depression, close to Thessaloniki (pop. about 1,000,000) was the site of a M$_s$ = 6.5 earthquake happened in June 1978. The second is the Kozani - Grevena (western Macedonia) area affected by the May 1995, M$_s$ = 6.6 shock. Both areas are dominated by typical active normal faults and the surface coseismic ruptures (of 1-10 cm displacement) formed along a main fault strand (~10 to 20 km respectively) and propagated on various other synthetic, antithetic, sympathetic and scissors faults. Main fault ruptures length are short in relation to released seismic moment, while fault segments and ruptured segments do not always match. The rupture pattern in both cases looks very similar. This enhances the hypothesis of multifractured type areas, which seem to dominate active tectonics in Greece. In such multifractured areas it is difficult to apply a particular segmentation model, since there seem to be many different fault strands quiescent during recent times, as shown by palaeoseismological trenching.

Keywords: Seismotectonics, Active tectonics of Greece, Thessaloniki earthquake, Kozani-Grevena earthquake, linkage active faults, geometrical barriers of faults.

INTRODUCTION

The irregular surface geometry of faults is strongly suggests that many fault systems are produced by complicated structure and fracture linkage on different scales [8, 47, 52]. A linked fault system can be defined as a network of broadly contemporaneous branching faults, which link-up over a length-scale much greater than individual fault segments. They are the most significant features of upper-crustal deformation, especially in highly deformed zones at high strain rates, like the Aegean, both at the margins of, and within, tectonic plates [8]. Earthquake ruptures often die out at barriers or asperities [1, 2, 17].

From the surface geological point of view, geometrical barriers are important in interpreting fault system geometry, linkage and kinematics. Geometrical barriers have been characterised as jogs (or Mode II) and steps (or bends or Mode III), depending on their orientation with respect to the slip direction [46, see also 16].

Although the surface traces of strike-slip faults are much straighter than dip-slip fault systems, an attempt will made here to recognise such structures in normal fault environments. Faults start as unlinked tension cracks, which eventually become linked by later shear cracks.

Seismic fault cracks usually follow a pre-existing fault zone deformed under a different stress (strain) regime, and it is very common to recognise strike-slip fault features in rejuvenated seismic faults as normal. Large fault zones over several seismic cycles are divided into segments on the basis of various criteria, including geometrical barriers [9, 50] that control rupture propagation. Individual fault segments between barriers regularly slip to produce of constant magnitude earthquakes (earthquake segments), that is characteristic earthquakes [48].

Palaeoseismological trenches and tectono-stratigraphic dating and interpreting fault displacements are also useful in fault segment division and defining characteristic earthquakes, or in testing alternative models [45].

The Aegean region is a typical area of active normal faulting in back-arc (southern Aegean) and intra - plate (northern) conditions. Some recent publications attempt to approach the problem of multi - fracturing in the region, based on non - uniform distribution of aftershock activity [10, 12, 21 and others] and structural heterogeneity [34, 43, 44]. Two multisegment seismogenic locations in northern Greece are examined in this paper, in respect mainly to their geometry, kinematics and segmentation problems, but also to their fault pattern similarities and differences (Fig. 1). The first is the Thessaloniki area, which suffered damaging seismic activity during the spring (24 May M_s=5.5) and summer (20 June M_s=6.5 and 3 August M_s=5.4) of 1978. During both major earthquakes manifestations of faulting were observed in the epicentral areas [4, 22-23, 30, 34].

The second is Kozani - Grevena, where on 13 May 1995 a M_s=6.6 earthquake occurred in a considerable 'aseismic' or 'low seismicity' area, where tectonic (seismic fault traces) and non tectonic origin (rock falls, landslides, liquefaction) phenomena were widespread over an 500 km^2 area [27, 40].

Both these areas are in northern continental Greece - Aegean broader region. The first lies in the Serbomacedonian geotectonic zone, between the Rhodope massif and the Axios (Vardar) zone, while the second stands on the Pelagonian-Sub-Pelagonian zones and Meso-Hellenic Trough.

The modern tectonic phenomena dominating the Aegean region are a N-S crustal extension and the western termination of the North Aegean right-lateral strike-slip fault, which accommodates most of the westward motion of the Anatolia [39]. Figure 1b shows the recent motion of the Aegean region [20, 29], where dashed lines are frown lines, and solid lines show isovelocity (5 mm/yr).

Although the Aegean is moving south-westward 4 cm/yr relative to Eurasia, the relative extension is ~1cm/yr. The whole Aegean region is deforming by 3 cm/yr in the northern part, 1-2 cm/yr in central continental Greece and 2 cm/yr in the south Aegean marginal sea. Along the north Anatolian - North Aegean Trough, slip rates vary from 0.6 cm/yr to 2 cm/yr (for references see [20]). The deformation observed in the region is complex and non homogeneous. A pure kinematic model of rigid plates or rigid blocks does not apply for the Aegean, where intra-plate (intracontinental) deformation is high, as the Thessaloniki 1978 and Kozani-Grevena 1995 earthquakes show.

THE THESSALONIKI AREA: GEOLOGICAL AND SEISMOLOGICAL BACKGROUND

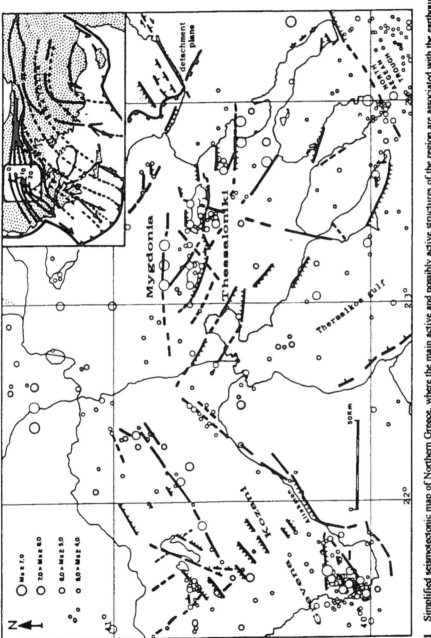

Fig. 1 Simplified seismotectonic map of Northern Greece, where the main active and possibly active structures of the region are associated with the earthquake epicentres (Thessaloniki-Mygdonia and Kozani-Grevena are the study areas). The top right hand box (figure 1b) shows the recent motion of the Aegean region [20] where dashed lines are flow lines and solid ones are isovelocity lines every 5mm/yr (insert is the northern Greece region-broader study area).

76   S. Pavlides et al.

Geology

The seismic sequence of June 20, 1978 (M_s=6.5) occurred in the Mygdonia depression (Volvi and Langada lakes), approximately 25 km northeast of the city of Thessaloniki (Macedonia, Greece), within the Serbomacedonian geological zone. Although the zone is a Palaeozoic and possibly pre-Palaeozoic crystalline massif with Mesozoic limestones and flysch (phyllites) deposits, affected mainly by the Alpide deformation, it is controlled by neotectonic and active faults as it is an integral part of the broader active Aegean region (Fig. 2).

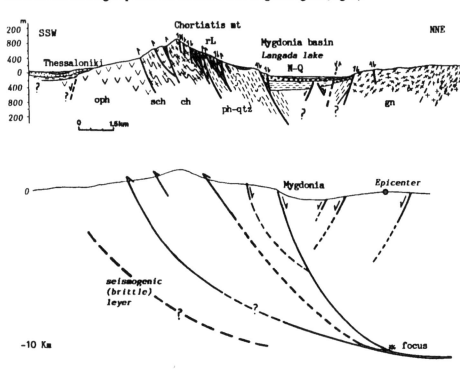

Fig. 2 a) A NNE-SSW trending cross-section through the Thessaloniki - Mygdonia seismogenic area (AA'), showing the mainly Alpide thrust structure and the post-Alpide high angle normal faulting. Position of AA' cross section is shown in figure 3. oph: ophiolites; sch: Jurassic green schists; ch: chert; rL: Triassic-Jurassic recrystallized limestones-marbles; ph-qtz: phyllites and quartzites; N-Q: Neogene-Quaternary sediments; gn: Palaeozoic gneisses, schists and amphibolites. b) A speculative sheme showing the possibly thrust Alpide pre - existing structure, which acts as a low angle plane and detachment surface at the brittle - ductile transition (~10-15 km), as well as the development of the neo - and active normal high angle faults.

Inherited structures which dominate the area are thrust and imbricate thrusts, reverse and strike - slip faults of the Alpide deformation, while post-orogenic extension created or reactivated normal or oblique slip faults. They are oriented primarily in a NW-SE and secondarily in N-S and NE-SW directions, while E - W trending structures are few and small, but orthogonal to the active stress regime of the region (Fig. 3).

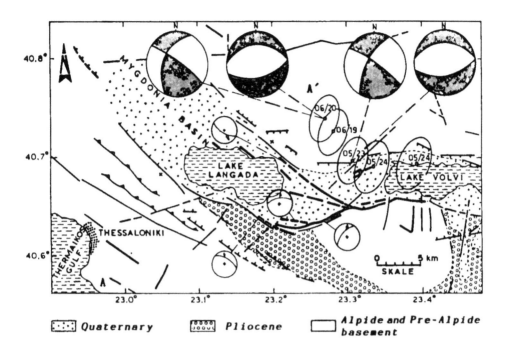

Fig. 3 A simplified sketch of the fault geometry and kinematics in the epicentral area of the 1978 (M_s=6.5) Thessaloniki earthquake (focal mechanisms and equal-area projection of representative fault and slip vectors as arrows). Thin lines are neotectonic faults (+footwall, -hanging wall) and thick ones the 1978 surface traces. Small arrows close to fauls indicate inferred sense of strike-slip displacements and discontinuous lines indicate faults of uncertain tectonic identity (AA': cross section of Fig. 2). [Ninety per cent confidence ellipse shown for each shock epicentre [4] focal mechanisms from [30] with predominantly strike-slip movement from short-period data; and [31], predominantly dip-slip from long-period data.

Destruction of an ocean basin during the late Cretaceous time and associated plate interaction caused regional-scale flexure, and Axios - Serbomacedonian suture. Rocks within the zone consist predominantly of deformed Mesozoic shelf carbonates, shales, and volcanocolastics metamorphosed to greenschist facies. The dominant metamorphism is considered to be late Cretaceous, while at least four main successive folding phases characterised the area from Cretaceous to Oligocene.

The neotectonic activity of the area forms a very active belt with shallow earthquakes of surface - wave magnitude up to M=7.0. The analysis and tectonic data shows that the epicentral distribution and the trend of isoseismals has the same orientation as well defined seismic zone trending NNW-SSE to WNW-ESE [36]. Five large, crustal shocks have occurred in this broader zone during this century (M>6.5).

The mean value of the fault slip rate inferred from seismicity for the zone, using the assumption that slip is not accommodated in viscoelastic process and the classical method of Davis and Brunne [7], has been calculated as 0.81 cm/yr [54], that is in good agreement with latest investigations [20]. Some segments of the fault zone show high slip rates of up to 1.05 cm/yr, some others less than 0.6 cm/yr, while individual segments of the active fault belts show unexpectedly low slip-rates of 0.03 to 0.06 mm/yr, as was found by tectonostratigraphic interpretation of palaeoseismological fault colluvium trenches [35].

The Thessaloniki seismogenic area is in fact a relatively small narrow Neogene-Quaternary basin, called the 'Mygdonia graben', filled with Neogene-Quaternary sediments. Geomorphic features are clearly exposed, especially along the most active segments of the marginal faults or within the basin (Fig. 3).

Seismotectonics

The main shock of June 20, 1978 was a double shock, small and large [18, 31]. They were based on several long and short period data recorded by seismographs and accelerographs with high magnification to show that a second event occurred 3 to 4 seconds after the first onset. That is the crack propagated from the main fault to a secondary linked segment, possibly of different orientation than the first one.

Fault plane solutions [4, 31, 51] of the main shock (1978, $M_s6.5$) and the largest fore-and after- shocks show E-W striking dip-slip normal faults. But the focal mechanisms, later revised by [30], and some of their results are remarkably close to the field data. They show that the seismic motion was mainly strike-slip sinistral along a NW-SE trending fault. Because the earthquake was a double event, the determined fault plane solution concerns the first of these events. The seismic fault is much longer than its trace observed extending northwest along the pre-existing structures. There are several other faults in the seismogenic area, some of which may have different properties from the main seismic fault. Additionally, it has been suggested that the focal mechanisms of small earthquakes six years after the 1978 sequence show a complex pattern [12]. The groups of faults as derived from their measurements fall into three categories; NW-SE trending planes with normal and sinistral component, dextral strike-slip NNE-SSW trending faults, and normal dip slip faults on approximately E-W trending planes.

Fore- and After-shocks distribution

The foreshock epicentres were mostly not distributed along an E-W direction, but followed an arcuate trend subparallel to and about 3 to 6 km north of the surface seismic ruptures. They were also concentrated in three distinct clusters [4, 51].

Surface geological research [23, 28, 36], including aerial photos and Landsat interpretation [38], shows that the faults which dominate the area can be divided into four distinct families: a) Faults striking N120° to N150°, with normal and sinistral strike - slip components. This direction shows excellent geomorphic features and constitutes the structural boundaries of the Langada sub-basin (western part of the Mygdonia Graben). b) Faults trending N10-20° with dextral strike-slip motion and no geomorphic features, c) Faults striking N40-60° and d) Structures striking between N80 and N100°, with essentially normal motion (dip-slip) (Fig. 3).

THE KOZANI-GREVENA AREA: GEOLOGICAL AND SEISMOTECTONIC SETTING.

Geology

The earthquake area lies in the western margin of the Pelagonian geotectonic zone to the Subpelagonian oceanic zone [3, 25]. Three main lithological groups can be distinguished in the region: The Pelagonian Alpide and pre-Alpide rocks, the sediments of the Meso-Hellenic Trough and the lacustrine-continental deposits of Plio-Quaternary age.

tectonic events took place during the last Alpide and Neotectonic periods. Two successive compressional tectonic events took place during the middle and late Miocene. The first, with an ENE-WSW direction of compression, caused NNW-SSE trending reverse faults, imbrication within the ophiolites and important right-lateral strike-slip faults trending E-W to NE-SW. The second event, with a NNE-SSW direction of compression, produced conjugate reverse faults trending E-W, and caused further imbrication of the ophiolites and the molassic sediments, including transfer strike - slip structures.

It is worth mentioning that the Cromion-Varis valley fault is a large right lateral strike-slip structure (Fig. 4), formed during the Miocene compressional deformation affects the ophiolitic complex in an E-W direction. This structure was detected by geophysical exploration (aeromagnetic research) carried out by the Institute of Geology and Mineral Exploration of Greece (I.G.M.E.) several years ago.

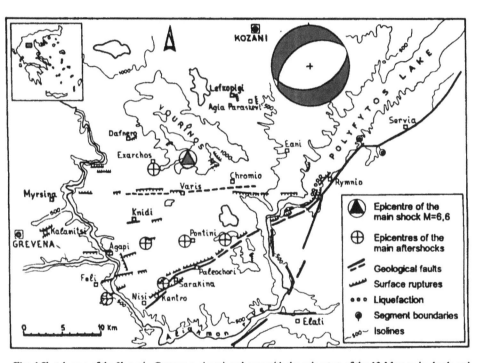

Fig. 4 Sketch map of the Kozani - Grevena meisoseismal area with the epicentres of the 13 May main shock and its major aftershocks. The principal geological faults (the Aliakmon fault and the Chromio - Varis Valley strike - slip fault) and the recent ground ruptures are shown. The liquefaction phenomena and the segment boundaries (geometrical barrier) between the Rymnio - Nisi and Servia segments (non - activated) are also shown. The sketch of the focal mechanism of the main shock was provided by the ERI automatic CMT inversion of the Geological Society of Japan [40].

An extensional post orogenic late Miocene-Pliocene tectonism followed the compressional deformation. This activated NNW-SSE trending normal faults mainly, as well as E-W and NE- SW trending pre-existing structures, and created the Ptolemais-Kozani basin. A subsequent middle Pleistocene extensional deformation rejuvenated or formed major normal faults trending NE-SW to ENE-WSW. This deformation is responsible for an atypical Basin and Range type of morphology that developed across these faults and appears to be still active. One of these structures is clearly visible on Landsat satellite images, with typical

geomorphic features revealed by a wealth of geological evidence, such as recent sediments
affected by the youngest faults, geomorphological features, and very recent striations on some
polished faults covered by unconsolidated recent alluvial fans [33, 37].

A Holocene reactivated structure which dominates the area is the impressive River Aliakmon
Fault (also known as the Servia fault). It is a typical impressive N70°E 30 km long dip-slip

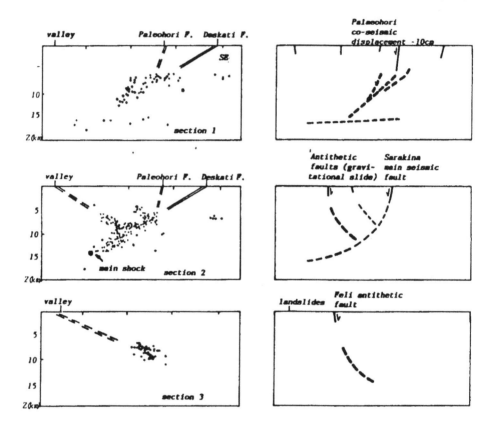

Fig. 5 Left: Cross sections perpendicular to the surface break showing the aftershock foci and the seismic fault
(double lines) as interpreted by [13]. Right: An alternative interpretation of the deep geometry of fault
distribution (listric or angular fault geometry of the main seismic fault and the antithetic structures). Heavy lines
are superficial high - angle normal faults.

normal fault, with a ~1,500 m displacement, composed of three segments, called Polyphytos,
Servia and Rymnio (Fig 6). Taking into account the Plio-Quaternary sedimentary deposits of
the hanging wall Kozani - Servia basin, the average slip rate has been estimated at 0.25
mm/yr with a coseismic slip of 0.5m and recurrence interval of about 2000 yrs [11], while 1-
2mm/yr, 2m slip per event , and a 1000-2000yrs time interval are possible [24].

Seismotectonics
A zone about 50 km long and 15-20 km wide was affected by 'faults', either reactivated
neotectonic ones, or other surface ruptures gravitational mainly and sympathetic faults,
during the spring-summer 1995 seismic sequence (Fig. 4). Surface fault traces directly
associated with the main seismic fault were observed in the area covered by ophiolitic rocks

and molasse type sediments between the villages of Palaeochori and Felli (Fig. 4) with a total length of about 12-15 km as well as in the unconsolidated recent sediments in the SW edge of the River Aliakmon fault (Rymnio). The ground ruptures consist of open fissures (1-10cm heaves) with small displacement (1-12 cm) of normal slip (dip towards NW). This rupture zone trending N75° coincides with the focal mechanism solution suggesting a fault striking N70°, dipping towards NW at high-angle, with a total length of 30 km. A series of subparallel antithetic ruptures with E-W strike, dipping towards the South, follows a second line (Chromio, Varis, Knidi) about 10 km long that may extend further west. Some of these ground ruptures are associated with this right-lateral strike - slip pre-existing fault affecting ophiolites mainly, which has been mapped by geophysical research and geological survey (I.G.M.E. see Fig. 4 and 6).

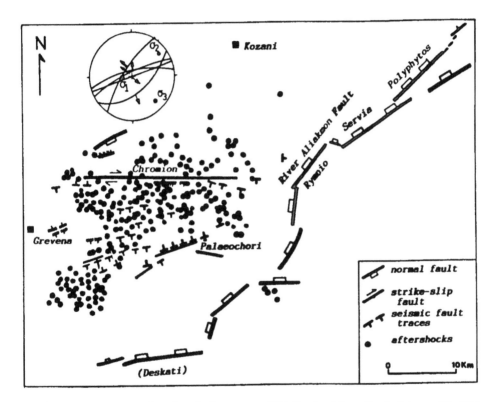

Fig. 6 The main structures of the Kozani - Grevena area (thick lines) and the surface fault traces of the May 1995, Ms=6.6 earthquake (stick thin lines) showing an 'imbricate fan' geometry along the WSW extension of the River Aliakmon active fault. Aftershock indicative distribution was taken from [13]. The seismic fault kinematics are a; so shown in a stereographic projection: faults as curves; slip-vectors as arrows (normal faults); σ_1, σ_2, σ_3 principal stress axes (NW-SE trending extension).

The previously calculated Quaternary and active stresses in the area [33, 37, 41] and the geometry of the active faults are in good agreement with the focal mechanism of the main shock.

The epicentres and hypocenters of the aftershocks are distributed mainly at the hanging wall, with the exception of SW cluster around the Nisi and Felli villages (Fig. 5, 6). They are associated to surface faulting, as well as other ruptures and ground failures (hanging wall block internal deformation) and occasionally at the foot wall, e.g. scissors secondary faults crossing the main seismogenic fault have been observed or assumed [24, 32, 40-41]. The SW edge (south of Grevena) of the activated area is dominated by the ENE-WSW trending and SE dipping antithtetic faults (Felli), as it is show by the aftershocks hypocenters (Fig. 5, 6) and surface measurements (strike N60-80° E, dip S80° E, pitch of striation S85° E).

It is worth noticing that although the total length of faulting has been calculated from seismological data to be of the order of 40 to 50 km [32], and the surface ruptures and related phenomena extend along a similarly sized zone, most of the seismic moment released from the main shock was associated with rupture of a restricted small segment named Palaeochori- Sarakina, as derived from the interferograms [24] and geodetic data [13].

The type and amount of displacement in 1995 was similar to a mid-Holocene event(s) documented on two palaeoseismological trenches [5, 35]. However the fault displacement of this palaeoevent suggests a multiple earthquake sequence.

The overall fault pattern of the area and the seismic ruptures geometry of the May-July 1995 earthquakes can be considered as a «horse tail» or «imbricate fan-splay faults» (Fig. 6). Although these terms refer to strike - slip structures, and in this case only dip - slip normal faults have been activated, they are borrowed because this geometry is similar to that of strike-slip pattern and second we believe that normal faults of the uppermost crust in the area follow pre-existing inherited structures, where strike-slip movements were dominant. The normal fault development in the area is Plio-Quaternary across the older structures, and the overall system is still developing, that is continuing maturing.

DISCUSSION AND CONCLUDING REMARKS

Both studied regions are characterised by thrusts, compressional imbricate structures, strike - slip and oblique slip faults of Alpide and post-Alpide deformation, as well as normal fault-bounded mountain or hill blocks. This later topography was formed by widespread extension during the past 5 My at least (10-15 Myr possibly) and mainly during middle-late Pleistocene and Holocene activity. The neotectonic stress regime has been described in detail in literature [23, 33, 36-37, 39, 41], while normal fault development by linkage of small structures has recently been shown [11, 34], for some Greek seismogenic faults, including the Thessaloniki one.

Once faults link up, displacement should be smaller on linked faults compared to an individual fault of equivalent width; and as deformation increases the irregular displacement distribution associated with linkages will be smoothed out [8, 42]. This is evident in the studied cases of the Thessaloniki (M_s=6.5) and Kozani-Grevena (Ms=6.6) earthquakes, where surface displacements (excluding gravitational phenomena) are extremely low (1-12 cm) in comparison with the worldwide average for earthquakes of equal magnitude. The phenomenon is repeated in continental Greece as we know from historical data and recent earthquakes (e.g. Atalanti - central Greece 1894, M_s=6.9-7.0; Thessaloniki 1902, M_s=6.6; Sophades-Thessaly 1954, M_s=7.0; Volos- Thessaly 1980, M_s=6.4; East Corinthian gulf 1981, M_s=6.7; Kalamata, South Peloponnese 1986 M_s=6.6), due mainly to linkage of relatively small multi - segmented faults [11, 34].

Another problem is the segmentation of active faults in continental Greece, as applied elsewhere [6, 9, 49-50]. Those earthquakes which produce surface ruptures in mainland Greece (M_s between 6.0 and 7.0) tend to have shallow hypocenters, commonly about 8 to 15 km deep (Thessaloniki 9 km; Kozani - Grevena 14 km) that is the seismogenic layer (brittle to ductile transition). They appear on the surface as 'geological structures' of a similar length (8-15 km), while seismogenic zones or rather seismic fault traces are consequently complex. Earthquake fault ruptures follow prominent faults (Gerakarou-Stivos in Thessaloniki area; (figures 3 and 7), Palaeochori-Sarakina-Felli in Kozani-Grevena case) (Fig. 4, 5), but local geometrical barriers usually deflect rupture propagation to varying degrees, producing multi-fractured seismogenic zones characterised by a diverse pattern and sense of fault movements [34], e.g. a main E-W trending seismic fault in Thessaloniki-Mygdonia controlled by NW-SE and NE-SW pre - existing structures, where rupture was propagated along an antithetic WNW-ENE oriented fault (Fig.7). An ENE-WSW trending and NNW dipping 15 to 20 km long fault accommodated the main deformation in the Kozani-Grevena event, while cross faults, small synthetic and antithetic ENE-WSW and E-W trending, SSE and S dipping structures were activated as well. Some possible underlying causes for the geometrical fault complexity of the study areas, as well as the widely distributed surface coseismic ruptures, which show great similarities with the active structures of the Basin and Range province in the western USA [9], could be the heterogeneous highly tectonized crust with pre-existing structures and fabrics, and varying lithologies, as well as important or governing right-lateral strike - slip structures (imbricate fans of fault termination on surface plan and possibly flower structures at depth). In such multifractured areas it is difficult to apply any one segmentation model. Instead we can segment complex fault zones as 'segment areas' or consider 'rupture zones'. The term 'rupture zones' was used to describe an area in which tectonic strain is substantially reduced by an earthquake [15]. But for large shallow earthquakes this can be extended to the deformed area, expressed mainly in the activated faults (seismic fault traces mainly and associated typical active faults possibly activated during the event) and other ground deformation. It can be defined by mapping surface faulting, liquefaction, rock falls, landslides, spring or thermal water changes, by the distribution of fore - and aftershocks and seismic intensity of a certain value (e.g. VII), by geodetic data showing coseismic deformation and finally by interferograms.

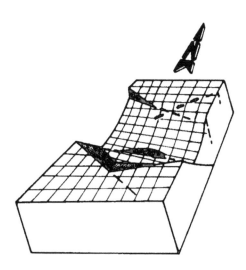

Fig. 7 Development of a hanging - wall relay E - W normal fault system at the convergence of two pre - existing oblique - slip structures. It is a simple model that may describe the active fault geometry in the Thessaloniki (Mygdonia) seismogenic area based on the detached hangingwall simple relay system proposed by [19, 43].

Each seismogenic area exhibits its own distinct structural characteristics and could interpreted with its own model, (e.g. Fig. 7 for Thessaloniki) but no single model is able to synthesise and explain all observations. Further data will increase our understanding of the structural complexities and earthquake rupture propagation. The structural style of extensional basins depends not only on the orientation of the controlling stress regime and the amount of extensional strain, but also on the availability of pre-existing upper crustal heterogeneity, lithological, bedding planes and mainly inherited structures (see also [14, 53]). In the case of the Thessaloniki and Kozani - Grevena areas, the N-S to NNW-SSE orientated extension (σ_3) has rejuvenated either orthogonal E-W or ENE-WSW trending pre-existing Alpide structures and oblique trending cross or scissor faults, parallel or sub-parallel synthetic and antithetic ones as secondary seismogenic structures. In both events their eastern linear fault strands remained inactive, while the aftershock epicentres and focal mechanisms (see [13, 32]) and the dispersal of seismic fault traces show a westward scatter pattern , like an imbricate fan or horse tail fault geometry.

REFERENCES

1. K. Aki. Characterization of barriers on an earthquake fault, *J. Geophys. Res.*, **84**, 6140 - 6148 (1979).
2. K. Aki. Asperities, barriers and characteristics of earthquakes, *J. Geophys. Res.*, **89**, 5867 - 5872. (1984).
3. J. H. Brunn. Contribution à l' étude géologique de Pinde septentrional et de la Macèdoine occidentale., *Ann. Geol. Pays Hellen.*, **7**, 1 - 358. (1956).
4. D. Carver and G. A. Bollinger. Aftershocks of the June 20, 1978, Greece earthquake: a multimode faulting sequence, *Tectonophysics*, **73**, 343 - 363 (1981).
5. A. Chatzipetros, S. Pavlides and D. Mountrakis. The 13 May 1995 Kozani - Grevena earthquake: a paleoseismological approach, *submitted to J. Geodyn.* (1996).
6. A. J. Crone and K. M. Haller. Segmentation and the coseismic behavior of Basin and Range normal faults: examples from east - central Idaho and southwestern Montana, U.S.A., *J. Struct. Geol.*, **13**, 151 - 164 (1991).
7. G. F. Davis and J. N. Brune. Regional and global fault slip rates from seismicity, *Nature*, **229**, 101 - 106 (1971).
8. I. Davison. Linked fault systems; Extensional, strike - slip and contractional. In: *Continental deformation*, P. Hancock (Ed.), Pergamon Press, 121 - 142 (1994).
9. C. M. dePolo, D. G. Clark, D. B. Slemmons and A. R. Ramelli. Historical surface faulting in the Basin and Range province, western North America: implications for fault segmentation, *J. Struct. Geol.*, **13**, 123 - 136 (1991).
10. K. Dimitropoulos and E. Lagios. Correlation between gravity observations and the aftershock sequence of the 11. Kalamata earthquake (13 - 9 - 1986) [in Greek with English abstract], *Bull. Geol. Soc. Greece*, **25**, 251 - 270 (1991).
12. T. Doutsos and I. Koukouvelas. Fractal analysis on normal faults in northwestern Aegean area, *Submitted to J. Geodyn. Spec. Issue (* (1996).
13. D. A. Hatzfeld, A. Christodoulou, E. M. Scordilis, D. Panagiotopoulos and P.M. Hatzidimitriou. A microearthquake study of the Mygdonian graben (northern Greece), *Earth Planet. Sci. Lett.*, **81**, 379 - 396 (1986/87).
14. D. Hatzfeld, V. Karakostas, M. Ziazia, G. Selvaggi, S. Leborgne, C. Berge, D. Diagourtas, I. Kassaras, I. Koutsikos, K.
15. Makropoulos, R. Guiguet, P. Paul, C. Papaioannou, R. Azzara, M. Di Bona, S. Baccheschi and P. Bernard. The Kozani - Grevena (Greece) earthquake of May 13, 1996, revisited from a detailed seismological study, *Bull. Seis. Soc. Am.* (in press).
16. J. J. Jarrige, 1992. Variation in extensional fault geometry related to heterogeneities within basement and sedimentary sequences. *Tectonophysics*, **215**, 161 - 166 (1992)
17. J. Kelleher. Rupture zones of large South American earthquakes and some predictions, *J. Geophys. Res.*, **77**, 2087 - 2103. (1972).
18. G. C. P. King. Speculations on the geometry of the initiation and termination processes of earthquake rupture and its relation to morphology and geological structure, *Pageoph.* **124**, 567 - 586 (1986).
19. G. King and J. Nabelek. Role of fault bends in the initiation and termination of earthquake rupture, *Science*, **228**, 987 (1985).
20. O. Kulhânek and K. Meyer. Spectral study of the June 20, 1978 Thessaloniki earthquake. In: *The Thessaloniki, northern Greece, earthquake of June 20, 1978 and its seismic sequence*. B. C. Papazachos and P. G. Carydis (Eds). pp. 187 - 199. Technical Chamber of Greece, Section of Central Macedonia, Thessaloniki (1983).
21. P.H. Larsen. Relay structures in a Lower Permian basement - involved extension system, East Greenland, *J. Struct. Geol.*, **10**, 3 - 8 (1988).
22. X. Le Pichon, N. Chamot - Rooke, S. Lallemant, R. Noomen and G. Veis. Geodetic determination of the kinematics of central Greece with respect to Europe: Implications for eastern Mediterranean tectonics, *J. Geophys. Res.*, **100**, 12675 - 12690. (1995).
23. H. Lyon - Caen, R. Armijo, J. Drakopoulos, J. Baskoutas, N. Delibasis, R. Gaulon, V. Kouskouna, J. Latoussakis, M. Makropoulos, P. Papadimitriou, D. Papanastasiou and G. Pedotti. The 1986 Kalamata (South Peloponnesus)

earthquake: Detailed study of a normal fault, evidences for east - west extension in the Hellenic arc, *J. Geophys. Res,.* 93, 14967 - 15000. (1988).

24. J. L. Mercier. Intra - plate deformation: a quantitative study of the faults activated by the 1978 Thessaloniki earthquakes, *Nature,* 278, 45 - 48 (1979).

25. J. L. Mercier and E. Carey - Gailhardis. Structural analysis of recent and active faults and regional state of stress in the Epicentral area of the 1978 Thessaloniki earthquakes (Northern Greece), *Tectonics,* 2, 577 - 600 (1983).

26. B. Meyer, R. Armijo, D. Massonnet, J. B. Chabalier, C. Delacourt, J. C. Ruegg, J. Achache, P. Briole and D. Papanastassiou. The 1995 Grevena (Northern Greece) Earthquake: Fault model constrained with tectonic observations and SAR interferometry, *Geophys. Res. Let.,* 23, 2677 - 2680 (1996).

27. D. Mountrakis. The Pelagonian zone in Greece: a polyphase deformed fragment of the Cimmerian continent and its role in the geotectonic evolution of East Mediterranean, *J. Geol.,* 94, 335 - 347 (1986).

28. D. Mountrakis, A. Kilias and N. Zouros. Kinematic analysis and tertiary evolution of the Pindos - Vourinos ophiolites (Epirus - Western Macedonia, Greece) (in press), *Bull. Geol. Soc. Greece* (1992).

29. D. Mountrakis, S. Pavlides, N. Zouros, A. Chatzipetros and D. Kostopoulos. The 13 May 1995 western Macedonia (Greece) earthquake. Preliminary results on the seismic fault geometry and kinematics. *XV Congress of the Carpatho - Balkan Geological Association,* September 17 - 20, 1995, Athens (in press).

30. D. Mountrakis, A. Psilovikos and B. Papazachos. The geotectonic regime of the 1978 Thessaloniki earthquakes. In: *The Thessaloniki, northern Greece, earthquake of June 20, 1978 and its seismic sequence.* B. C. Papazachos and P. G. Carydis (Eds). pp. 11 - 27. Technical Chamber of Greece, Section of Central Macedonia, Thessaloniki (1983).

31. B. C. Papazachos, A. A. Kiratzi and E. E. Papadimitriou. Orientation of active faulting in the Aegean and surrounding area, *6th Congress of the Geol. Soc. of Greece* (in press) (1992).

32. B. C. Papazachos, D. Mountrakis, A. Psilovikos and G. Leventakis. Surface fault traces and fault plane solutions of the May - June 1978 major shocks in the Thessaloniki area, Greece, *Tectonophysics,* 53, 171 - 183 (1979a).

33. B. Papazachos, D. Mountrakis, A. Psilovikos and G. Leventakis. Focal properties of the 1978 earthquakes in the Thessaloniki area, *Bulgarian Geophys. J.,* 6, 72 - 80 (1979b).

34. B. C. Papazachos, D. G. Panagiotopoulos, E. M. Scordilis, G. F. Karakaisis, C. Papaioannou, B. G. Karacostas, E. E. Papadimitriou, A. A. Kiratzi, P. M. Hatzidimitriou, G. N. Leventakis, P. S. Voidomatis, K. I. Peftitselis, A. Savaidis and T. M. Tsapanos. Focal properties of the 13 May 1995 Large (Ms=6.6) Earthquake in the Kozani Area (North Greece). *XV Congress of the Carpatho - Balkan Geological Association,* September 17 - 20, 1995, Athens (in press).

35. S. Pavlides. Neotectonic evolution of the Florina - Vegoritis - Ptolemais basin (W. Macedonia, Greece). *Ph.D. Thesis, University of Thessaloniki,* pp. 265 (in Greek) (1985).

36. S. Pavlides. Active faulting in multi - fractured seismogenic areas; examples from Greece, *Z. Geomorph. N.F.* 94, 57 - 72. (1993).

37. S. Pavlides. First palaeoseismological results from Greece. *Annali di Geofisica,* 34, 545 - 555 (1996).

38. S. B. Pavlides and A. A. Kilias. Neotectonic and active faults along the Serbomacedonian zone (SE Chalkidiki, northern Greece), *Ann. Tectonicae,* 6, 97 - 104 (1987).

39. S. Pavlides and D. Mountrakis. Extensional tectonics of northwestern Macedonia, Greece, since the late Miocene. *J. Struct. Geol.* 9, 4, 385 - 392. (1987).

40. S. Pavlides and N. Soulakellis (1991). Multifractured seismogenic area of Thessaloniki 1978 earthquake (Northern Greece). In: *Proceedings of International Earth Sciences Congress on Aegean Regions, Izmir, Turkey. 1 - 6 October 1990,* M.Y. Savaşcin and A.H. Eronat (Eds), 2, 64 - 75.

41. S. Pavlides, D. Mountrakis, A. Kilias and M. Tranos. The role of strike - slip movements in the extensional area of Northern Aegean (Greece). A case of transtensional tectonics, *Ann. Tectonicae,* 4, 197 - 211 (1990).

42. S. B. Pavlides, N. C. Zouros, A. A. Chatzipetros, D. S. Kostopoulos and D. M. Mountrakis. The 13 May 1995 western Macedonia, Greece (Kozani Grevena) earthquake; preliminary results, *Terra Nova,* 7, 544 - 549 (1995).

43. S. Pavlides, D. Mountrakis, A. Chatzipetros, N. Zouros and D. Kostopoulos. The Grevena - Kozani (May 13, 1995) earthquake, western Macedonia, Greece: seismogenic faulting in an «aseismic» area. In: *3rd Workshop Statistical models and Methods in Seismology ESC,* G. Papadopoulos (Ed), Proceedings (in press) (1996).

44. D. C. P. Peacock and D. Sanderson. Displacements, segments linkage and relay ramps in normal fault zones, *J. Struct. Geol.,* 13, 721 - 733 (1991).

45. G. Poulimenos and T. Doutsos. Barriers on seismogenic faults in central Greece, *J. Geodynamics,* 22, No. 1/2, pp. 119-135, 1996.

46. G. P. Roberts and I. Koukouvelas. Structural and seismological segmentation of the gulf of Corinth fault system: implications for models of fault growth, *Annali di Geofisica,* 34, 619 - 646 (1996).

47. C. H. Scholz. Comments on models of earthquake recurrence. In: *Fault segmentation and controls of rupture initiation and termination,* D. Schwartz and B. Sibson (Eds), *U.S.G.S. Open - File Rep.* 89 - 315, 350 - 360 (1989)

48. C. H. Scholz. The mechanics of earthquakes and faulting. Cambridge University Press, pp. 449 (1990).

49. C. H. Scholz, C. A. Aviles and S. G. Wesnousky. Scaling differences between large interplate and intraplate earthquakes, *Bull. seism. Soc. Am.,* 76, 65 - 71 (1986).

50. D. P. Schwatz and K. J. Coppersmith. Fault behaviour and characteristic earthquakes: examples from the Wasatch and

San Andreas fault zones, *J. Geophys. Res.*, **89**, 5681 - 5698 (1984).

51. D. P. Schwartz and K. J. Coppersmith. Seismic hazards: New trends in analysis using geologic data. In: *Active Tectonics (studies in Geophysics)*, National Academy Press, Washington, 215 - 230 (1986).

52. D. B. Slemmons and C. M. dePolo. Evaluation of active faulting and associated hazards. In: *Active Tectonics (studies in Geophysics)*, National Academy Press, Washington, 45 - 62 (1986).

53. C. Soufleris, J. A. Jackson, G. C. P. King, C. P. Spencer and C.H. Scholz. The 1978 earthquake sequence near Thessaloniki (northern Greece), *Geophys. J. Royal Astron. Soc.*, **68**, 429 - 458 (1982).

54. J. S. Tchalenko, 1970. Similarities between shear zones of different magnitudes. *Bull. Geol. Soc. Am.*, **81**, 1625 - 1640.

55. V. Tron and J. P. Brun. Experiments on oblique rifling in brittle - ductile systems, *Tectonophysics*, **188**, 71 - 84.

56. P. S. Voidomatis, S. B Pavlides and G. A. Papadopoulos. Active deformation and seismic potential in the Serbomacedonian zone, northern Greece, *Tectonophysics*, **179**, 1 - 9 (1990).

Proc. 30th Int'l. Geol. Congr., Vol. 5 pp. 87-99
Ye Hong (Ed)
© VSP 1997

Geomorphic Evidence for Irregular Faulting along the North Piedmont of Liulengshan Range, Shanxi Rift System, China

XU XIWEI[1), DENG QIDONG[1), WANG YIPENG[1), YONEKURA NOBUYUKI[2) AND SUZUKI YASUHIRO[2)
1)Institute of Geology, State Seismological Bureau, Beijing 100029,China
2)Department of Geography, University of Tokyo, Tokyo 113, Japan

Abstract

Three basic geomorphic units, which have developed in response to the climate changes during the last glaciation, are recognized along the northern piedmont of the ENE-trending Liulengshan Range from airphoto interpretation and field observation. These three units are the inter-stadial fluvial fan (S3) about 23-53 ka old, the maximum-glacial fluvial fan (S2) about 10-23 ka old and the post-glacial fluvial fan (S1) about 0-10 ka old, respectively. Five fluvial terraces (T0-T4) have also developed owing to the climatic temperature flucturation on thousand-years scale. These geomorphic units and fluvial terraces are chronometrically dated by thermoluminescence dating method. Striped geomorphic mapping shows a natural segmentation in offset geomorphology along the north Liulengshan fault. A set of quantitative data on faulting age, vertical throw and slip rate are obtained from long-distance topographic profile levellings across the offset geomorphic units or fluvial terraces. The average Holocene vertical slip rates of the north Liulengshan fault are in the range of 0.43-0.55 mm/a. Late Quaternary irregular faulting is demonstrated by temporal and spatial ebbing-and-flowing variation in vertical slip rate. The period of 1.56-7.6 ka BP is the main phase of intensely vertical faulting along the northern piedmont of Liulengshan Range. Paleoseismic data show that its surface-rupturing earthquake recurrence interval reaches 5,000-6,000 years. The coseismic vertical throw for the latest one reaches (3.4±0.3) m, and its elapse time has reached 6,000-7,200 years on the Huajialing-Xiejiayao segment, which indicates that the north Liulengshan fault is now in the risk of surface-rupturing earthquake occurrence.

Keywords: Offset-Geomorphology, Irregular Faulting, Quaternary, Shanxi Rift System

INTRODUCTION

Tectonic base level fall partly determines the relief of a fluvial system, and a pulse of tectonic uplift along a mountain front will cause adjustment in the fluvial system that flows across the front [1]. Differential vertical motion is one of many variables that affect configuration of the hillslope, stream, and depositional subsystem of an arid fluvial system. The effect of pulsatory uplift of mountains relative to the adjacent basins induced from normal-faulting slip results in a distinctive suite of landform elements: multiphase peneplanation surfaces in the mountains, stream-erosional or stream-depositional terraces across the mountain piedmont, and fluvial floodplains or fans in the basins [2, 20]. Those landform elements of different orders virtually guarantees temporal variations in rates of erosion and sedimentation owing to enviromental changes in localized tectonic process and regional climate [1, 6, 7, 10]. Tectonic process commonly persists over millions of years and profoundly affects landscape of mountain ranges and subsidence basins. Fluvial system, in contrast, responds in tens to hundreds of years to disruption of what otherwise is equilibrium [6, 12], which lead to alternating styles of landform development on time scales of thousands of years [7]. Quantitative characteristics of offset-geomorphology (ages of geomorphic surfaces and associated offset throws) along a fault trace has objectively recorded the faulting process, motion amount and spatial distribution, thus it is very useful in analysing the faulting behaviour.

This paper presents a spatial geometry of late Quaternary normal faults, basic geomorphic units, and offset geomorphic surfaces by stripped geomorphic mapping along the northern piedmont of Liulengshan Range or the north Liulengshan fault (NLF), from airphoto interpretation and field observation. Furthermore, the basic geomorphic uints and offset geomorphic surfaces are chronologized by determining the stratigraphic age of the lowest superjacent cover above the alluvial fills, using thermoluminescence (TL), infrared stimulated luminescence (IRSL) or Carbon 14 dating methods. The fault throws are measurred by long-distance topographic profile levelling using Auto-Level Meter. Trench technique across fresh fault scarp is also used in order to obtain paleoseimic data and to recover faulting process, which is essential for assessing seismic hazard on an active fault.

GEOLOGICAL SETTING

The north Liulengshan fault (NLF) is the most southern major-bordering fault of the Yangyuan sub-basin, located at the northern piedmont of the E-W trending Liulengshan Range in the northern extensional domain of the Shanxi rift system, China (inset map in Fig. 1). The fault is typical of dip-slip normal fault about 130 km long. It strikes N70°-80°E, dips toward NNW with a dip angle of 60-80°, and has controlled the development of the Yangyuan sub-basin [19, 20]. Together with a series of ENE-trending normal faults at the north Henshan, north Wutaishan and north Xizhoushan piedmonts, the NLF has controlled

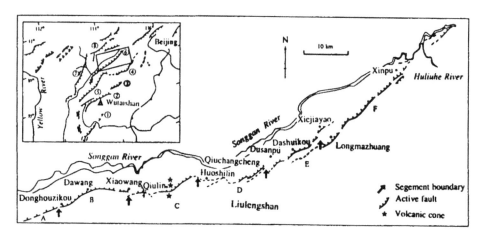

Fig.1 Sketch map showing fault trace and segmentation along the north Liulengshan fault (NLF). (1) north Xizhoushan F.; (2) north Wutaishan F.; (3) north Taibaiweishan F.; (4) south Yuxian F.; (5) north Hengshan F.; (6) north Liulengshan F.; (7) Kouqiu-Emaokou F.; and (8) Yanggao-Tianzhen F.

the basic pattern of the basin-and-range structure in the northern extensional domain of the Shanxi rift system [13, 16]. Previous studies indicate that the Holocene vertical slip rates of the north Xizhoushan and north Wutaishan faults reach 0.76-1.2 mm/a, about 0.8 mm/a for the north Hengshan fault, 0.34-0.57 mm/a for the north Liulengshan and north Taibaiweishan faults as well as the south Yuxian fault [16-18]. Paleomagnetic data from the Tertiary basalts show that the blocks of Hengshan and Liulengshan Ranges have undergone tilting toward S190-200°W and their maximum tilting angles have reached 11°and 6° since the Pliocene, respectively [17].

The northern extensional domain, where the north Liulengshang fault is located, is one of the places along the Shanxi rift system where destructive historical earthquakes have often struck. The 512 AD Daixian earthquake (M=7.5), 1038 AD Xinxian-Dingxiang earthquake (M=7.25), 1626 AD Lingqiu earthquake (M=7) and 1668 AD Yuanping earthquake (M=7) ruptured the north Wutaishan, north Xizhoushan and north Taibaiweishan faults and the Yunzhongshan fault, respectively [15, 19]. Although no surface rupturing earthquakes (M≥7) have occurred along the south Yuxian, north Hengshan and north Liulengshan faults, yet six historical earthquakes (M≥5) have struck the Yu-Guang basin which is mainly controlled by the south Yuxian fault on its southern border, including the 1581 AD (M=6) and the 1618 AD earthquakes (M=6.5). Two earthquake swarms (M_s=6.1 and M_s=5.8) successively occurred along the north Liulengshan fault in the Yangyuan sub-basin in 1989 and 1991, respectively, which are considered to be closely associated with abrupt motion of the north Liulengshan fault [14].

GEOMORPHIC FEATURES

Methods
Determination of recent Quantitative slip history of an active fault requires recognition of active fault traces and offset geomorphic features along with description, analysis, comparison and dating of the superjacent loess/paleosol stratigraphy deposits on the various geomorphic surfaces. A critical aspect of this method need to recognize offset topographic features across the fault scarps and also geomorphic units covering the fault traces to establish slip sense of recent surface ruptures. Airphoto interpretation to determine the basic geomorphic units for stripped geomorphic mapping is an economic and effective approach to this problem. The northern piedmont of the ENE-trending Liulengshan Range was geomorphically mapped at a scale of 1:50,000 by airphoto interpretation and field investigation (Fig. 2). Geomorphic features mapped include fault traces and scarps, stream channels, displaced fluvial fans, and other surfaces. Large-scale geomorphic mapping is done at key place where offset geomorphic units or offset-terraces are well preserved, and long-distance topographic profile levelling across the offset geomorphic units (surfaces) is also made with a distance meter (Auto-Level) in order to obtain a set of quantitative data (faulting age, fault throw, Holocene slip rate and paleoearthquake sequence).

Basic Geomorphic Units and Chronology
The most southern border of the ENE-trending Yangyuan sub-basin is marked by an abrupt front of Precambrian metamorphic rocks and Paleozoic limestones produced by normal slip along the north Liulengshan fault. Coarse clastic debris shed off this range front and from the interior of the range has been deposited as alluvial or fluvial fans along the northern margin of Liulengshan Range. Coarse-grained alluvial deposits grade abruptly basinward into fined-grained deposits. Several ages of fluvial deposits and their associated surfaces are represented by fluvial fans and terraces at different geomorphic levels (Fig.2).

In most cases fined-grained loess-like deposited cover has developed above the top of fluvial fills (basic geomorphic units). According to the previous study on sedimentary origin of those loess-like covers overlying various geomorphic surfaces in the loess region of central China, loess-like dustfall is so persistent that any newly created geomorphic surface immediately begins to accumulate a cover of wind-blown (eolian) sediment [9]. Therefore. by determing its stratigraphic age of the lowest loess or paleosol above a fluvial fill, the stratigraphic age of the fluvium can be inferred. Determination of the absolute age of the displaced geomorphic surface was based on the dated superjacent loesse or paleosol at top of terrace or flood-plain fluvium. That is, the age for underlying fluvium must be equivalent to at least that of the lowest part of the superjacent loess or paleosol above a fluvial fill. Age samples for those geomorphic surfaces are collected in the field and are dated by thermoluminescence (TL) dating method at Institute of Geology, State Seismological Bureau, China.

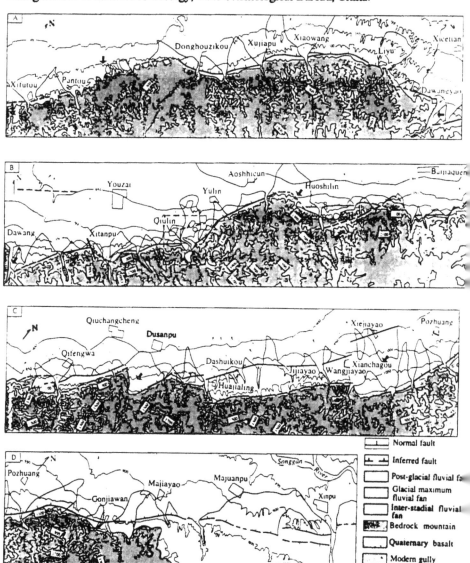

Fig. 2 Detailled geomorphic map along the north Liulengshan fault (NLF)

Three basic geomorphic surfaces, which have developed in response to climate change during the last glaciation, are recognized along the northern piedmont of the ENE-trending Liulengshan Range by detailled airphoto interpretation, field investigation along the active fault traces, structural analysis of superjacent loess or paleosol, and TL sample collection and dating. Those three units or geomorphic surfaces (fluvial fills) are the inter-stadial fluvial fan (S3) about 53-23 ka old, the maximum-glacial fluvial fan (S2) about 23-10 ka old, and the post-glacial fluvial fan (S1) about 10-0 ka old, respectively, which correlate with glacial stages. Whereas several terraces have also developed during inter-glacial stages [20].

The post-glacial fluvial fan (S1) further includes the modern fluvial fan, youngest terrace (T0) and early Holocene terrace (T1). The flood-plain gravels (alluvial fill) for T0 are directly exposed on the ground surface that may correlate broadly with episodes of neoglaciation during the past 5,000 years that are recognized globally [4]. Two-layer-structure of alluvial fill (gravel layer) and a cover of wind-blown loess about 0.5-0.7 m thick has developed for T1 terrace. The TL age of the lowest part of the covering loess is dated to be (7.6±0.6) ka BP at south of Dusanpu Village and to be (7.4±0.6) ka BP at Dashuikou Village, and this age may be a little younger than that of the underlying alluvial fills.

The maximum-glacial fluvial fan (S2) also includes two fluvial terraces (T2 & T3) with two-layer-structure of the flood-plain gravel layer on the lower and a cover layer of wind-blown loess on the top. The TL ages of the lowest part of the covering loess are dated to be between 10.3-13.5 ka BP for T2 terrace and 15.6-18.7 ka BP for T3 terrace.

The inter-stadial fluvial fan (S3) consists of a series of small gully-net incised alluvium, which correspond to the highest terrace (T4) observed along the northern piedmont of Liulengshan Range. Its superjacent cover belongs to the late Pleistocene Malan loess. It is worth to point out that there is a 15-m-thick medium- to fine-grained sand layer of fluviolacustrine facies unconformably overlying the Precambrian metamorphic rocks, but underlying the inter-stadial fluvial fills between Wangjiayao and Jijiayao Villages. The TL age of the top part of the fluviolacustrine facies is dated to be (109.6±9.2) ka BP, while the bottom of the superjacent Malan loess for T4 terrace, which is about 0.2-0.5 m thick, to be (43.7±3.4) ka BP between Wangjiayao and Jijiayao Villages, (48.1±3.9) ka BP at Huajialing and (48.4±3.9) ka BP at south of Dusanpu Village.

The above three basic geomorphic units, which may respond to the climate changes within the last glacial period, can be taken to be the basic units for stripped geomorphic mapping along the northern piedmont of Liulengshan Range. Five orders of fluvial terraces can be identified at some places, which can be used for local large-scale geomorphic mapping in detail and long-distance topographic profile levelling across the offset geomorphic surfaces to obtain quantitative slip data.

Geomorphic Segmentation Along Fault Traces
Stripped geomorphic mapping shows a natural geomorphic segmentation along the northern Liulengshan Range (north Liulengshan fault) owing to significant variation in geomorphic features, including offset geomorphology, discontinuous fault traces on the ground surface, relative fault scarp heights and evolutional history of micro-physiographic stages. Six segments can be recognized in the mapping areas and they are the Xifutou, Donghouzikou-Dawangcun, Qiulin, Huoshilin-Qiuchangcheng, Huajialing-Xiejiayao and Longmazhuang-Xinpu segments from the west to the east, respectively (Fig. 1 & 2). Those segments differ from each other in offset geomorphology and faulting behaviour.

The Xifutou segment (A) is characterized several subparallel step-like faults [14]. These faults had cut the inter-stadial fluvial fans and late Pleistocene Malan loess, which are dated to be 53-23 ka BP, to form fault scarps about 20-25 m high and are covered with glacial maximum and post-glacial fluvial fans. The latter two are dated to be 23-10 ka BP and 10-0 ka BP, respectively [20]. Thus, the latest faulting on the Xifutou segment took place before 23 ka BP.

The Donghouzikou-Dawangcun segment consists of a single fault trace lying between the Quaternary loose deposits and Pre-Cenozoic bedrocks. The fault trace is continuously outcroped and has vertically offset interstadial-postglacial fluvial fans to form fault scarps with different heights. The highest scarp reaches 20-30 m, the top of which is covered with Malan loess. The bottom of the Malan loess above the alluvial fill is dated to be (28.6 ± 2.3) ka BP. The latest faulting has vertically offset the 7.4-7.6-ka-old deposits, but is covered with the modern fluvial fan 0-5 ka BP.

Three phases of basaltic eruptions have occurred in late Quaternary along the Qiulin segment. The oldest basalt is characterized by a lot of vertical joints filled the Qiulin Gully on the foot-wall side and its K-Ar age is dated to be (1.71 ± 0.4) Ma BP. The younger one is distributed on both the foot-wall and Hanging-wall sides of the north Liulengshan fault and covers the earlier basalt. Its TL age of the optalic metamorphic layer between the pre-Cambrian rock and basalt is dated to be (182.9 ± 13.9)-(197.6 ± 15.4) ka BP. The multiple faulting has vertically offset this phase-basalt to form step-like scarps with small graben structure. Topographic profile across the Liulengshan fault shows that the fault throw has reached 65-70 m, which yields a long-term vertical slip rate of 0.36 mm/a on the Qiulin segment. The basalt of latest eruption covers the older basalts at the piedmont of Liulengshan Range, but no age data are available. Besides, a set of left-lateral normal faults have developed west of Xitanpu Village with a NW strike, SW dip and a dip angle of 55-70_. Those faults have cut the pre-Cambrian metamorphic rocks and late Pleistocene sub-sandy soil which is dated to be $(21,540\pm260)$ years BP (^{14}C age).

The postglacial fluvial fans have well developed and few of the inter-stadial and glacial maximum fluvial fans are outcroped along the Huoshilin-Qiuchangcheng segment. The fault has cut the inter-stadial and glacial maximum fluvial fans to form discontinuous fault scarps and the fault traces are covered with   postglacial fluvial fans.

The Huajialing-Xiejiayao segment consists of two step-like small faults. The basin-range bordering fault (F_1) at the most southern margin of the sub-basin is covered with the Malan loess (43.7 ± 3.4)-(48.4 ± 3.9) ka old, indicating that this fault has become inactive since late Pleistocene. The active fault (F_2) has migrated basinward 1.7-3.5 km far from the basin-range bordering fault (F_1). F_2 fault consists of en-echelon discontinuous normal faults, and they have vertically offset fluvial fans of different ages. The latest faulting has cut the terrace (T1) 7.6-7.4 ka BP to form low fault scarps 0.5-3.4 m high. The fault traces are covered at some places with a surface-soil layer, the bottom of which is about (2.2 ± 0.14) ka old (TL age).

The Longmazhuang-Xinpu segment consists of a single continuous fault, which is located along the northern piedmont of Liulengshan Range. The polished bedrock fault plane is observed at many places. The fault strikes N45°-60°E and dips toward the NW with a dip angle of 65°-85°. The striation rakes on the bedrock fault planes are 77°-90°, showing that the fault is dominated by   normal components. The fault has systematically offset the glacial-maximum fluvial fans to form 8-10 m high fault scarps, or offset the inter-stadial fluvial fans to form 35-50 m high fault scarps.

QUANTITATIVE OFFSET GEOMORPHOLOGY

Offset Geomorphic Features

Through detailled geomorphic mapping, chronology for geomorphic units and long-distance topographic profile levelling by using Auto-Level meter across the offset geomorphic units or fluvial terraces, vertical throws in given temporal periods are obtained on the Donghouzikou-Dawangcun, Huajialing-Xiejiayao and Longmazhuang-Xinpu segments, and this provides a basis to reconstruct faulting processes on those fault segments.

On the Donghouzikou-Dawangcun segment the fault has vertically offset the inter-stadial, maximum-glacial and post-glacial fans (σ_1, σ_2 & σ_3) and four terraces (T1-T4), but only T_2 terrace can be found on both the hanging- and foot-walls of the north Liulengshan fault. Long-distance topographic profiles crossing the fault scarps show that the vertical throws of T2 terrace, which is dated to be (12.9 ± 1.0)-(13.4 ± 1.1) ka BP, reach 5.5-6.4 m at Donghouzikou and Xiaowang Villages (Fig. 3a), yielding a vertical slip rate of 0.43-0.48 mm/a (Table 1).

Table 1. Vertical fault throws and average slip rates of the north Liulengshan fault

Segment Name	Location	Vertical Throw (m)	Terrace No. and Their TL ages(ka)	Average Slip Rate (mm/a)
Donghouzikou-	Donghouzikou	0.55	T0: ?	
Dawangcun	Donghouzikou	5.5±0.4	T2: 12.9±1.0	0.43
	Xiaowang	6.4±0.6	T2: 13.4±1.1	0.48
	Xiaowang	14±2	T3: 15.6±1.2	0.95
Huajialing-	Dusanpu	0	T0: ?	
Xiejiayao	Dusanpu	F1: 2.2±0.2	T1: 7.6±0.6	
	Dusanpu	F2: 1.2±0.5		
	Dusanpu	Total: 3.4±0.3		0.45
	Dusanpu	F1: 3.7±0.1	T2: 12.8±1.0	
	Dusanpu	F2: 4.9±0.4		
	Dusanpu	Total: 8.6±0.4		0.67
	Dusanpu	F1: 4.2±0.2	T3: 18.7±1.5	
	Dusanpu	F2: 4.6±0.2		
	Dusanpu	Total: 8.8±0.4		>0.47*
	Dusanpu	21±0.9	T4: 48.4±3.9	0.43
	Jijiayao	23±2	T4: 43.7±3.4	0.53
	Dashuikou	>0.5±0.2	T1: 7.4±0.6	>0.1
	Dashuikou	7.7±0.2	T2: 10.3±0.8	0.75
	Xiejiayao	5.9±0.4	T2: 13.5±1.1	0.43
Longmazhuang-	Longmazhuang	8.2±0.1	T3: 15.6±1.3	0.53
Xinpu	Gongjiayao	9.7±0.2	T3: 17.7±1.4	0.55

*It is known that there was a quiet period during 18.7~12.8 ka BP on the Huajialing- Xiejiayao segment.

The Huajialing-Xiejiayao segment is the best one where the offset geomorphic surfaces has been well preserved. Local geomorphic mapping there shows that north Liulengshan fault consists of two subparallel normal faults (F1 & F2 in Fig. 4). Those two faults have repeatedly offset various fluvial fans and terraces since the late Pleistocene, but are covered with modern fluvial fan or youngest terrace (T0). Their fault scarps and vertical throws of corresponding geomorphic surfaces increase as their ages become older (Table 1). Long-distance profile-levellings crossing the fault scarps further show that the two faults have not cut the terrace T_0 about 0-3 ka old, since the terrace T_0 keeps its original topographic slop (Fig. 3b & 4), but the terraces T1- T4. Thus the Holocene vertical slip rate is obtained to be 0.45-0.67 mm/a. Similarly, the vertical slip rates can also be obtained from profile levelling at Jijiayao, Dashuikou and Xiejiayao Villages. The average slip rate has been 0.43-0.53 mm/a since 40-50 ka BP, 0.43-0.75 mm/a since 10-13.5 ka BP and 0.45 mm/a since 7.4-7.6 ka BP (Table 1).

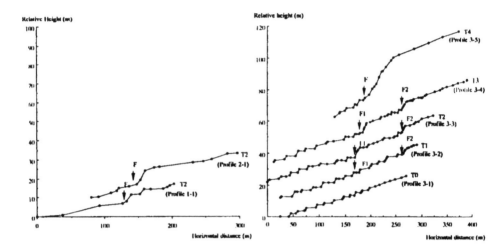

Fig. 3 Topographic profiles of offset terrace (T0-T4) on the Donghouzikou-Dawangcun segment (a) and the Huajialing-Xiejiayao segment south of Dusanpu Village (b). Profile 2-1: Donghouzikou Village; Profile 1-1: Xiaowang Village.

Although multi-faultings have produced fault scarps with different heights, only the vertical throws of T3 terrace can be measured to be 8.2-9.7 m at Longmazhuang and Gongjiayao Villages through long-distance profile levelling, which gives an average vertical slip rate of 0.55-0.53 mm/a since 15.6-17.7 ka BP (Table 1).

Irregular Faulting Behaviour

The above data show that the north Liulengshan fault has been active with a Holocene vertical slip rate of 0.43-0.75 mm/a. More detailed chronology of the T_1-T_4 terraces and their corresponding throws obtained from profile levelling makes it possible to reconstruct the faulting sequence on the Huajialing-Xiejiayao segment south of Dosanpu Village (Figs. 3b & 4). Taking the vertical throw of 3.4 m for the terrace T_1 as the latest surface-faulting event (one paleoearthquake), this event took place during the past several thousands of years before T_0 formation, but after T_1 formation. The silimar vertical throws of 8.6-8.8 m for T_2 and T_3 terraces may include two or three events after T2 terrace formation (about 7.6 ka BP) and a 6-ka-quiescence during the period for T_2 and T_3 formation (between 7.6 ka BP and 12.8 ka BP). The vertical throw of 21±0.9 m for T_4 terrace may have be accumulated during the past several events. Thus, the faulting process along the Huajialin-Xiejiayao segment is uneven and irregular in late Quaternary, and this is demonstrated by temporal ebbing-and-flowing flucturation in vertical slip rate (Fig.5). The period of 12.8 to 7.6 ka BP is the most intense phase of vertical faulting along the northern piedmont of the Liulengshan Range.

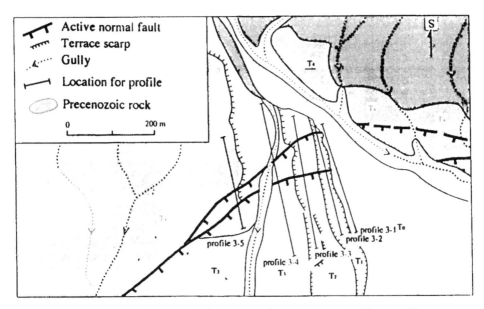

Fig.4   Detailled offset geomorphic map on the Huajialing-Xiejiayao segment south of Dusanpu Village

HOLOCENE PALEOSEISMICITY

No surface-rupturing earthquakes have recorded along the north Liulengshan fault since at least 1500 years BP. Thus no destructive earthquake catalogue is available for us to analyse the future seismic risk for this active fault. Trenching technique has expanded to address problem of paleoearthquake faulting, folding, ground failure, and faulting-induced sedimentation, and this technique is useful to establish a Holocene surface-rupturing earthquake catalogue for active faults. Two trenches have been excavated across the fault scarps on the T1 and T3 terraces on the Huajialing-Xiejiayao segment south of Dusanpu Village, respectively (Fig. 4).

Geologic Log of T1-1 Trench   The T1-1 trench is excavated across the fault scarp on the T1 terrace. The trench strikes NW about 8 m long, 2 m wide and 2.5 m deep. Its stratigraphic sequence exposed is described below with the same number order as shown in Fig. 6a.
① Red-brown pebble layers intercalated with sub-sandy soil layer. There are several individual large gravels marked by shadowed areas in the pebble layers.
② Grayish-yellow gravel layer with loose sedimentary structure.
③ Earthy-yellow subsandy soil layer which is dated by IRSL (Infrared Stimulated Luminescence) to be (6.7±0.5) ka BP. This soil layer is bent owing to the latest normal slip on the fault (F2).
④ Earthy-yellow gravels forming a colluvial wedge, which represents a paleoearthquake event.

⑤ Earthy-yellow subsandy soil layers containing gravels. Two IRSL samples from the top and bottom of those soil layers give ages of (6.6±0.5) ka BP and (5.0±0.3) ka BP, respectively.

⑥ Imbricated sandy gravels in the fault zone (F1).

⑦ Earthy-yellow sub-sandy soil forming a filling wedge and this filling wedge represents a paleoearthquake event. The IRSL age of the bottom part of the filling wedge is dated to be (6.6±0.6) ka BP.

⑧ Gray-yellow surface soil layer with sands and gravels.

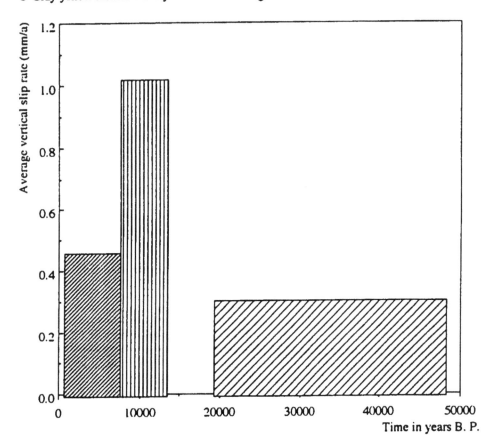

Fig. 5   Flucturation in time vs vertical slip rate on the Huajialing-Xiejiayao segment south of Dosanpu Village

In the trench two fault branches (F1 & F2) have been revealled. The F1 has cut the layer ③, but covered with layer ⑤. Between the layer ③ and layer ⑤ is the colluvial wedge ④. F2 has cut the layer ① and covered with the filling wedge ⑦. According to the above relationship between the fault branches and stratigraphic sequences exposed in the trench log as well as their chronology, we can know that the latest paleoearthquake took place at (6.6±0.6)-(6.7±0.5) ka BP. In other words, only one surface-rupturing earthqauke occurred after T1 terrace formation on the Huajialin-Xiejiayao segment.

Geologic Log of T3-1 Trench   The T3-1 trench is excavated across the fault scarp on the T3 terrace. The trench strikes NW about 10 m long, 2 m wide and 3.5 m deep. Its stratigraphic sequence exposed in the trench is described below with the same number order as shown in Fig. 6b.

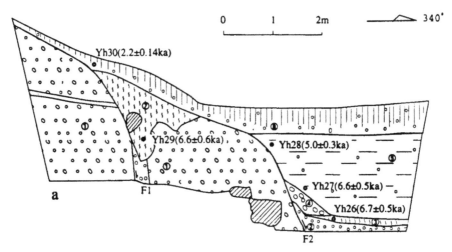

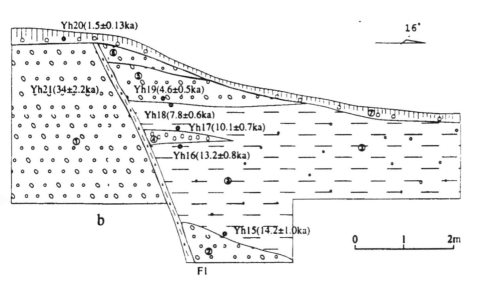

Fig. 6 T1-1 and T3-1 Trench logs across the fresh fault scarps on T1 and T3 terraces south of Dusanpu Village. Location is shown in Figure 4.

① Red-brown gravel layers with fined-grained sands. Its IRSL age on its top reaches (34±2.2) ka BP.

② Grayish-yellow coarse gravel layer with a colluvial wedge structure, which may represents a paleoearthquake-related sediments.

③ Earthy-yellow sub-sandy soil layers with gravels. Four samples taken from its bottom, middle and top give their IRSL ages in the range of (7.8±0.6)-(14.2±1.0) ka BP.

④ Grayish-yellow fine gravel layer representing a small colluvial wedge associated with a paleoearthquake.

⑤ Grayish-yellow gravels with a loose structure forming a colluvial wedge. The IRSL age of its bottom part is dated to be (4.6±0.5) ka BP.

⑥ Gray and grayish-yellow sandy gravel layer.

⑦ Gray-yellow surface soil layer containing sands and gravels. Its IRSL age is about (1.5±0.13) ka BP.

The fault exposed on the trench log has cut the layers ①-⑥, but is covered with the surface soil layer ⑦ Three colluvial wedges on the hanging wall represent three paleoearthquakes. According to the IRSL ages dated, it is know that those three paleoearthquakes occurred at (4.6±0.5)-(7.8±0.5) ka BP, (10.1±0.7)-(13.2±0.8) ka BP, and before (14.2±1.0) ka BP, respectively.

From the above trenches, three surface rupturing paleoearthquake events can be identified. They occurred at (6.6±0.6) ka BP, 11.7±2.4 ka BP, and before (14.2±1.0) ka BP. Thus the surface rupturing earthquake recurrence interval on the Huajialin-Xiejiayao segment reaches 5,000-6,000 years. The coseismic throw for the latest event   reaches (3.4±0.3) m in terms of data from long-distance profile levelling across the fault scarp on the T1 terrace. Furthermore, since the elapse time for the latest paleoearthquake reaches 6,000-7,200 years on the Huajialin-Xiejiayao segment, which has already surpassed its recurrence interval, the north Liulengshan fault is estimated to be in the risk of surface rupturing earthquake occurrence in the future.

CONCLUSIONS

Three basic geomorphic surfaces and five terraces have developed along the northern Liulengshan Range. Irregular multi-faultings have vertically offset various geomorphic units along the different segments of the north Liulengshan fault. The average Holocene vertical slip rates are 0.43-0.48 mm/a for the Donghouzikou-Dawangcun segment, 0.45 mm/a for the Qiulin segment, and 0.53-0.55 mm/a for the Longmazhuang-Xinpu segment. The period of 15.6-7.6 ka BP is the main phase of intensely vertical faulting along the northern piedmont of Liulengshan Range. The paleoseismic data further shows that the north Liulengshan fault is in the risk of surface rupturing earthquake occurrence in the future.

Acknowledgements

The authors are very grateful to Seismological Science Foundation of China (contract no. 95103) and Japan Science Foundation (Monbusho contract no. 05041054) for financial supports. This is the contribution No. 970001 of Institute of Geology, State Seismological Bureau, China.

REFERENCES

1. W. B. Bull, L. D. McFadden. Tectonic geomorphology north and south of the Gariock fault, California. In: *Geomorphodogy in arid regions.* Doehring, D. O.(Ed.). Annual Binghamton Conference, Binshamton, New York, P. 115-138 (1977).
2. W. B. Bull. Stream-terrace genesis: implications for soil development, *Geomorphology* 3, 351-367 (1990).
3. A. J. Crone, K M. Haller. Segmentation and the seismic behavior of basin and range normal faults: Examples from east central Idaho and southwestern Montana, U. S. A., *J. Struct. Geol.* 13, 151-164 (1991).
4. J. M. Grove. *The Little Ice Age,* Methuen, London (1988).
5. G. Kukla, Z. S. An, J. L. Melice, J. Garin, and J. L. Tiao. Magnetic Susceptibility record of Chinese Loess, *Transactions of the Royal Society of Edinburgh (Earth Seiences)* 81, 263-288 (1990).
6. D. Merritts, K R. Vincent. Geomorphic response of coastal stream to low, intermediate, and high rates of uplift, Mendocino triple junction region, northern California, *Geol. Soci. Amer. Bull.* 101, 1373-1388 (1989).

7. P. Molnar, E. T. Raisbeck, B. C. Brown et al. Quaternary climate change and the formation of river terraces across growing anticlines on the north flank of the Tien shan, China. *J Geol* 102, 583-602 (1994).

8. K. J. Mueller, T. K. Rorkwell. Late Quaternary activity of the Laguma Salada fault in northern Baja California, Mexico. *GSA Bull.* 107, 8-18 (1995).

9. S. C. Porter, Z. S. An, H. Zheng. Cyclic Quaternary alluviation and terracing in a nonglaciated drainage basin on the north flank of Qinling shan, central China. *Quaternary Research 38*, 157-169 (1992).

10. S. A. Schumn. Alluvial river response to active tectonics. in: Wallace, R., panel chairman, *Active tectonics: Studies in Geophysics Series*, Gtophysics Research Forum, National Academy Press, P. 80-94 (1986).

11. Wei B. Wang K, Yao Z. The 1989 Datog-Yanggao earthguake: focal mechanism and seismotectonic implications, *Earthquake Research in China* 8, 199-208 (1994).

12. Williams G, M G Wolman. Downstream effects of dams on alluvial rivers. *U. S. Geolosical Survey professional paper* 1286, Washington, 1-83 (1984).

13. Xu X. W., and Deng Q. D. The basin and range structure in the tensile area in the northern part of Shanxi Province and the mechanism of formation, *Earthquake Research in China* 4, 19-27 (1988). (in Chinese with English abstract)

14. Xu X. W., Y. Che, Z. Yang, H. You, Y. Wang, and Y. Zhu. Discussion of the seismogenic structure model for the Datong-Yanggao earthquake swarm, *Earthquake Research in China*, 8, 361-371 (1992).

15. Xu X. W., Q. D. Deng, R. Dong, C. Zhang, and W. Gao. Study on strong earthquake activity and risk areas in the Shanxi graben system, *Seismology and Geology*, 14, 305-316 (1992). (in Chinese with English abstract)

16. Xu X W, Ma X Y. Geodynamics of the Shanxi rift system, China. *Tectonophysics* 208, 325-340 (1992).

17. Xu X W, X Y Ma and Q D Deng. Neotectonics of the Shanxi rift system, China. *Annales Tectonicae*, special Issue, Supplemtnt to Volame VI, 40-53 (1992).

18. Xu X W, X Y Ma and Q D Deng. Neotectonic activity along the Shanxi rift system, China, *Tectonophysics* 219, 305-325 (1993).

19. Xu X W, Ma X Y, Deng Q D, Liu G D, and Ma Z J. *Neotectonics, Paleoseismology and Ground Fissures of the Shanxi Rift System, China. 30th IGC field trip guide T314*. Geological Publishing House, Beijing, China, 152pp (1996).

20. Xu X. W., Yonekura N., Suzuki Y., Deng Q., Wang Y., Takeuchi A., and Wang C. Y. Geomorphic study on late Quaternary irregular faulting along the northrn piedmont of Liulengshan Range, Shanxi Province, China, *Seismology and Geology* 18, 169-181 (1996).

Proc. 30ᵗʰ Int'l. Geol. Congr., Vol. 5 pp. 101-114
Ye Hong (Ed)
© VSP 1997

Seismic Potential along the Main Active Strike-slip Fault Zone in Western Sichuan, China

WEN XUEZE

Seismological Bureau of Sichuan Province, Chengdu, Sichuan 610041, China

Abstract

The 900-km-long main strike-slip fault zone of western Sichuan has been exhibiting strongly left-lateral faulting. This is one of the major seismogenic belts in southwestern China. This paper combines geologic and historic-earthquake information to evaluate quantitatively seismic potential along this fault zone. Average slip-rates along the fault zone have been recalculated or reestimated. Based on fault geometry and space-time pattern of historical earthquake ruptures, this fault zone has been divided into 16 segments. Combining estimated co-seismic slips, historic and prehistoric event timings, as well as using the time-predictable and renewal models, the author estimates the average recurrence time between earthquakes for each fault segment. Further, the probabilities of future segment-rupture earthquakes have been computed by employing the evaluating model of time-dependent probabilistic seismic hazard. The main results show as follows: (1) Till A.D. 2026, 6 out of the 16 segments have relatively high cumulative probabilities (>0.45). These 6 segments are all located at the seismic gaps that have been not ruptured for at least 100 years.(2) In the coming 30 years (1996-2026), not all these 6 segments have relatively high conditional probabilities, because most of them have longer average recurrence times or shorter elapsed times relative to the average recurrence times. (3) On the basis of comparisons among the probabilities of the individual segments, it is suggested that two areas, from Mianning to Kangding (segments 8 to 11) and from Shimian to Xichang (segments 14 to 15), should be considered as relatively risk area along the fault zone in the coming years.

Keywords: seismic potential, active fault, Western Sichuan

INTRODUCTION

The studied fault zone is extending through the whole western Sichuan from the northwest to the southeast (Fig. 1), which has a total length of about 900 km, and is made up of four faults, the Ganzi-Yushu fault, the Xianshuihe fault, the Anninghe fault and the Zemuhe fault. Since the late Quaternary, this fault zone has been exhibiting strongly left-lateral faulting[4]. Also this fault zone is one of the main seismogenic belts of southwestern China. Since the 18th century, at least 20 earthquakes with magnitude 6.5 and greater have occurred along the fault zone.

This paper tries to evaluate quantitatively seismic potential along different segments of the fault zone, based on data of fault slip-rate, paleo-earthquake and historical earthquake, and on the time-dependent probabilistic model.

Figure 2 is a flow-chart to present the research procedure of this paper. This procedure is similar with that has been used by some researchers or research groups and carried out on active plate boundaries[9,19,20,21].The author of this paper has been making efforts to apply such a procedure to some of intra-plate active faults on the China's mainland[15,16,17,18]. This paper is one of these efforts.

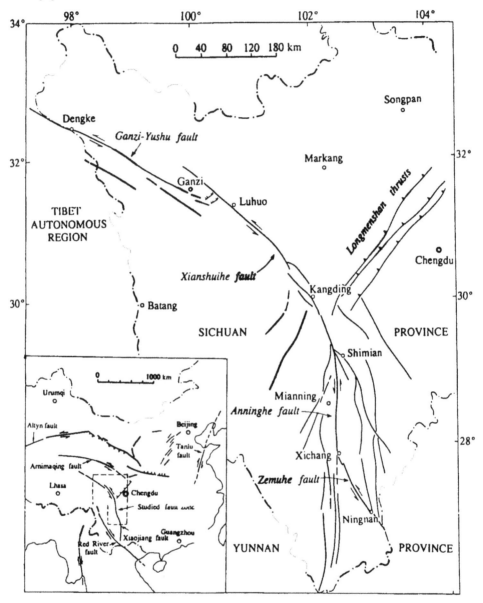

Fig. 1. Map showing the main strike-slip fault zone in western Sichuan. The inset map shows regional relationship between the studied fault zone and other major active faults of the China's mainland.

AVERAGE FAULT SLIP RATE

Although some estimates about slip-rates of the studied fault zone have been published since the 1980s, differences still remain in these estimates. The author of this paper has analyzed carefully the data reported by the previous researchers[6,11,13,14,18,22], then, recalculates or reestimates the average slip-rates and their standard deviations. The recalculated or reestimated average slip-rates are shown in Figure 3. For 9 localities in Figure 3, where reliable measurements of geomorphic offsets and dates of sediment have been available, recalculated average slip-rates and their errors are obtained. For other 3 localities in Figure 3, average slip-rates and their uncertainties (figures in brackets) are referred reasonably.

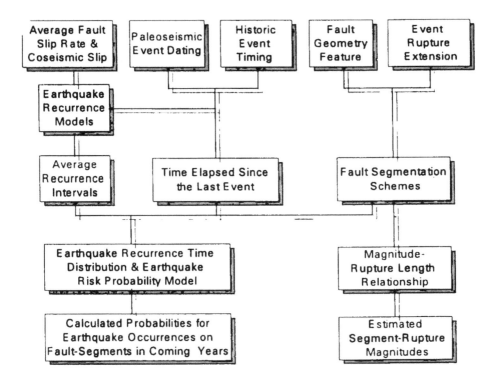

Fig. 2. A flow-chart showing procedure adopted in this paper for estimating quantitatively earthquake potential on active faults.

Figure 3. shows that the Ganzi-Yushu and Xianshuihe faults have relatively high slip-rates that are as much as 10 to 14 mm/yr. However, along the Anninghe and Zemuhe faults, slip-rates are only between 5.5 and 6.5 mm/yr. It is suggested from Figure 1 that there are more secondary fault branches around the Anninghe and Zemuhe faults. A reasonable explanation seems to be: these secondary branches share the horizontal movement components of the faulted-blocks, and hence reduce the slip-rates along the major traces of the Anninghe and Zemuhe faults.

HISTORIC EVENTS AND THEIR RUPTURE SPACE-TIME PATTERN

Historical earthquake data are available for the studied fault zone, excepting for two fault sections. Figure 4 is a set of maps showing spatial distribution of the historical earthquake sources, drawn in 5 different periods from the early 18th century to the present. The dimensions of these sources are delimited on the basis of the areas of severely damaged zones during the earthquakes.

During the last 250 years, historic events repeated for twice or three times along the fault section near Luhuo and Daofu (Fig.4). This section just has the highest slip-rate (between 13 and 14 mm/yr.) of the fault zone(Fig.3). So, along the fault zone, the higher slip-rate is, the higher earthquake recurrence rate is. If taking these source lengths as corresponding rupture ones, and making them become function of time, we get a space-time pattern of the historic ruptures (Fig. 5).

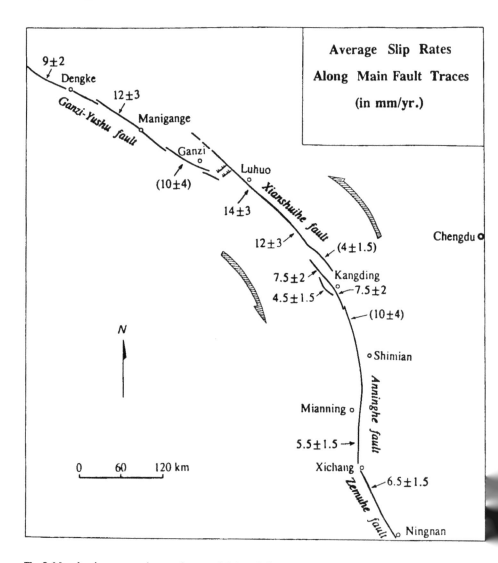

Fig. 3. Map showing recent estimates of average left-lateral slip-rates along the studied fault zone.

Figure 5 suggests:

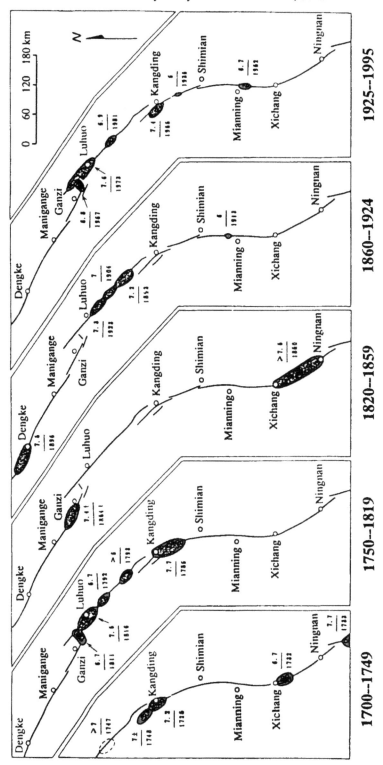

Fig. 4. Maps showing spatial distribution of historical earthquake sources along the studied fault zone, drawn in 5 different periods from early 18th century to the present. The source dimensions are delimited according to severely damaged areas during the earthquakes.

(1) Along the fault section near Manigange, there is a space-time domain without documented earthquake record, which means that no information about historically documented events can be obtained for the period before the 18th century there, excepting one event (occurred round about A.D. 1506) has been dated roughly by archaeological method[17]. However, along the fault section between Mianning and Xichang, a 500-year-long earthquake history is available.

(2) The historic ruptures tend to repeat at the previous localities, but the amount of spatialoverlaps between adjoining ruptures is smaller relative to their rupture lengths.

Three seismic gaps with different lengths are able to be identified along the fault zone. These gaps have undergone time periods of at least 100 years without segment-rupture earthquakes since the last events.

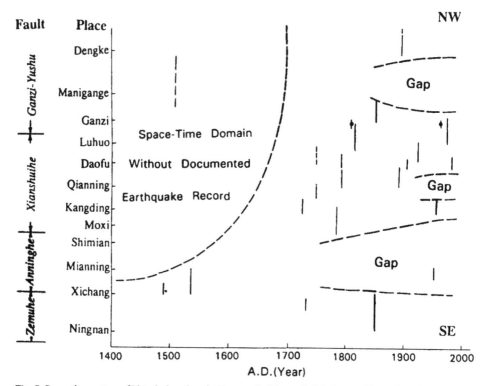

Fig. 5. Space-time pattern of historical earthquake "ruptures" of the studied fault zone. The ordinate shows spatial localities along the fault zone's strike from southeast to northwest. The dashed vertical-lines show uncertain rupture extensions.

FAULT SEGMENTATION

To segment this fault zone is to divide it into relatively independent rupture units. In determining segment-boundaries, the follows are taken into account: (1) Larger-scale geometric discontinuance along the fault zone, such as step-overs between en echelon faults or shorter fault branches (Lasting boundary). (2) Joint between adjoining fault sections that have repeated more than one historical ruptures (Relatively stable boundary). (3) Connection between fault sections that have ruptured only once historically (Uncertain boundary). (4) Ends of the longest historic rupture, if a fault section took place earthquakes with variable rupture lengths (Uncertain boundary).

Figure 6 presents the segmentation model for the studied fault zone, in which 16 segments have been divided and marked with S1, S2, S3,....., respectively.

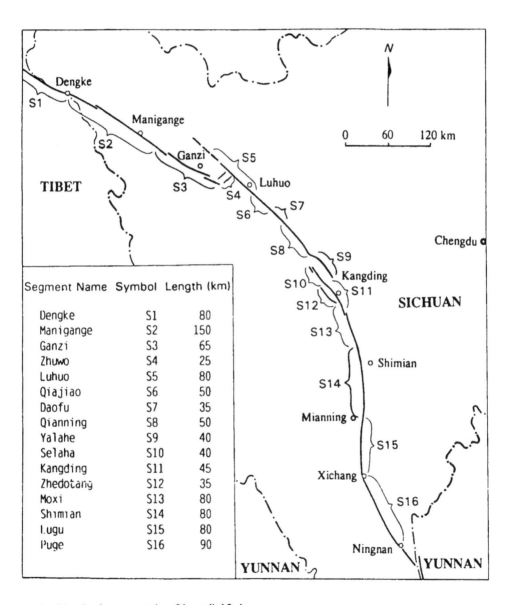

Segment Name	Symbol	Length (km)
Dengke	S1	80
Manigange	S2	150
Ganzi	S3	65
Zhuwo	S4	25
Luhuo	S5	80
Qiajiao	S6	50
Daofu	S7	35
Qianning	S8	50
Yalahe	S9	40
Selaha	S10	40
Kangding	S11	45
Zhedotang	S12	35
Moxi	S13	80
Shimian	S14	80
Lugu	S15	80
Puge	S16	90

Fig. 6. Map showing segmentation of the studied fault zone

EARTHQUAKE AVERAGE RECURRENCE INTERVAL

The methods used to estimate average recurrence interval between earthquakes are as follows:

For a fault-segment where the average slip-rate, v, and the mean coseismic-slip, u, of the last event are available or able to be estimated, the average recurrence interval, Tm, will be calculated on the basis of both the "time-predictable model"[11] and the "renewal model"[21].

For the time-predictable model:

$$T_m = \frac{u}{v} \tag{1}$$

$$\sigma_d^2 = \left(\frac{S_u}{u}\right)^2 + \left(\frac{S_v}{v}\right)^2 \tag{2}$$

Where
T_m: Average (or median) recurrence interval;
σ_d, : Data-uncertainty of T_m;
u: Mean coseismic slip of the last event;
S_u: Standard deviation of u;
v: Average fault slip-rate(not including creep-rate);
S_v: Standard deviation of v.
For the renewal model:

$$T_m = \frac{u_m}{v} \tag{3}$$

$$\sigma_d^2 = \left(\frac{S_u}{u_m\sqrt{n}}\right)^2 + \left(\frac{\sigma_I}{\sqrt{n}}\right)^2 + \left(\frac{S_v}{v}\right)^2 \tag{4}$$

Where
u_m: Mean of n coseismic slips;
S_u: Standard deviation of u_m;
n: Number of previous earthquake slips. When only the last slip is available, n=1;
σ_I: Intrinsic-uncertainty of recurrence interval [see equation (10)].
For most fault-segments, the mean coseismic-slip, u, is estimated by such a procedure[18]: inputting the surface-wave magnitude, M, the rupture length, L, and the maximum coseismic-slip, Dmax, into a group of empirical relationships between (u·L) and M, and among u, M, L and Dmax , we get several estimates for the mean coseismic-slip of the last event. Then, assuming the weight of each estimate is inversely proportional to the variation of the estimate, take the weighted mean as the best estimate. One out of the 16 fault segments, the Shimian segment(S14), has available ^{14}C dating ages for the previous 4 paleoseismic events[10]. Based on these ages, the author of this paper has recalculated the average (or median) recurrence time and its data-uncertainty from[9,18]

$$T_m = \exp[\ln T_{av} - (\mu + 0.5\sigma_I^2] \tag{5}$$

$$\sigma_{4} = \sqrt{\dfrac{1}{\displaystyle\sum_{i=1}^{n} \dfrac{1}{(S_{i}/T_{i})^{2}+\sigma_{I}^{2}}}} \qquad (6)$$

Where

T_{av}: Arithmetic mean of recurrence intervals of n events;

μ: Mean of lognormal distribution of recurrence intervals [(see equation (9)];

T: Median of time interval between events;

S: Standard deviation of T, it will be zero once events are historical;

n: Number of interval between events.

For the 16 fault segments, the data u9ed to calculate average recurrence intervals and their uncertainties have been summarized in Table 1. Along with the elapsed times since the last events, the calculated average recurrence intervals be used to compute the occurrence probabilities of segment-rupture earthquakes.

Table 1. Data base for calculating average recurrence intervals of 16 fault segments

Fault Segment		Fault Aver-age Slip Rate	Historic or Prehistoric Events		Data for the Last Event		
					Rupt. Length	Max. Slip	Mean Slip
Name	Symbol	(mm/yr.)	Date	Mag.	(km)	(m)	(m)
Dengke	(S1)	9±2	1896	7.5	80	5	2.30±0.4
Manigange	(S2)	12±3	1506±70	8.1	150	13	5.10±0.95
Ganzi	(S3)	10±4	1854±	7.4	60		2.35±0.7
Zhurwo	(S4)		1811	6.7			
			1967	6.8			
Luhuo	(S5)	14±3	1816	7.5±			
			1973	7.6	83	3.6	2.40±0.4
Qiajiao	(S6)	13±4	1747	>6.7			
			1923	7.3	55	3	1.90±0.35
Daofu	(S7)	13±4	1792	6.7			
			1904	7			
			1981	6.9	36		1.25±0.3
Qianning	(S8)	12±3	1793	>6			
			1893	7.2	53	3.5	1.70±0.3
Yalahe	(S9)	4±1.5	1700?	7	41		1.35±0.3
Selaha	(S10)	7.5±2	1748	7	40		1.45±0.45
Kangding	(S11)	7.5±2	1725	7.2	50		2.00±0.6
Zhedotang	(S12)	4.5±1.5	1955	7.4	35	3.1	2.95±0.55
Moxi	(S13)	10±4	1786	7.7	90		3.00±0.7
Shimian	(S14)		1030±90	>7			
Lugu	(S15)	5.5±1.5	1489	6.7			
			1536	7.5			
			1952	6.7			0.45±0.10
Puge	(S16)	6.5±1.5	1732	6.7			
			1850	7.5	90	8	3.30±0.6

CALCULATION OF SEGMENT-RUPTURE PROBABILITIES

The approach of evaluatine5 long-term seismic potential on the individual fault segments is based upon a model of earthquake occurrence that assumes that the probability of an earthquake along a fault segment increases with elapsed time From the last event there. Such s model is so called the time-dependent probability model[9,21]. This paper has computed two kinds of probabilities, conditional probability and cumulative probability.

Conditional probability, P_e, is likelihood of an earthquake occurring in the time interval from T_e to T_e and DT, conditional on the earthquake not having occurred prior to T_e:

$$P_e\left(T_e < T \le T_e + DT|T > T_e\right) = \frac{\int_0^{T_e + DT} f(T)dT - \int_0^{T_e} f(T)dT}{1 - \int_0^{T_e} f(T)dT} \tag{7}$$

Where, $f(T)$ is a probability density of random recurrence intervals, T. The zero is a relative starting-point of time, and is set up at the date of the last earthquake. T_e is a period from the date of the last earthquake to January 1 st, 1996. DT is a time interval predicted, which has been taken as long as 30 years.

Cumulative probability, F, is likelihood of an earthquake occurring in the period from the last event time to T_e and DT:

$$F\left(T < T_e + DT\right) = \int_{T_e}^{T_e + DT} f(T)dT \tag{8}$$

$f(T)$ has been assumed as a lognormal density function. In this paper, the generic distribution of recurrence time for characteristic earthquakes, which has the density function[8,9],

$$f(T) = \frac{1}{T \cdot \sigma_N \cdot \sqrt{2\pi}} \exp\left\{\frac{-\left[\ln(T/T_m) - \mu\right]^2}{2\sigma_N^2}\right\} \tag{9}$$

has been adopted. Where m (=-0.01) is the mean of this lognormal distribution. The net uncertainty, σ_N, is consisted of two components, the data uncertainty, σ_d, and the intrinsic uncertainty, σ_I.

$$\sigma_N = \sqrt{\sigma_d^2 + \sigma_I^2} \tag{10}$$

The data uncertainty, σ_d, comes from the uncertainty of the estimated average recurrence interval, T_m. The intrinsic uncertainty, σ_I (=0.21), comes from the generic distribution.

MAGNITUDE ESTIMATE OF FUTURE SEGMENT RUPTURE EARTHQUAKE

For strike-slip fault-segments, the magnitudes of the future segment-rupture earthquakes have been roughly estimated by a selected group of empirical relationships :

$M = 6.24 + 0.619 \log L$,	for worldwide [2]	(11-1)
$M = 4.94 + 1.296 \log L$,	for USA and China [2]	(11-2)
$M = 5.92 + 0.880 \log L$,	for Qing-Zang Plateau[5]	(11-3)
$M = 5.12 + 0.579 \ln L$	for SW China[18]	(11-4)

Where, L is in kilometers. Assuming a coming characteristic event ruptures a corresponding;

segment length, L, four estimates for the event magnitude can be obtained from these 4 formulas. Their mean will -be taken as the best estimate of the coming magnitude.

S4, the Zhuwo segment, is the only one non-strike-slip segment of the all 16 segments (see Fig. 6), which is located at the southeastern edges of the Ganzi pull-apart area. This fault segment exhibited NE-trending normal faulting during the 1967 earthquake of magnitude 6.8[1,18]. The coming magnitude of this segment is thus estimated by the relationship for worldwide normal fault earthquakes[12]:

$$M = 0.809 + 1.34 \log L(m) \qquad (12)$$

PROBAILISTIC SEISMIC POTENTIAL ANALYSIS

Table 2 lists the calculated probabilities and the predicted characteristic magnitudes for the future segment-rupture earthquakes. Since two kinds of recurrence models, the time-predictable and renewal models, have been used, and they give somewhat different probability values, the mean probabilities of these two models have been taken as the final result.

Table 2. Calculated conditional probabilities, P, (1996-2026), and cumulative probabilities, F,(till A. D. 2026), for earthquake occurrences on the 16 fault-segments

Fault Segment		Length	Segment-rupture Time-Pred. M.		Event Renewal M.		Probabilities Mean		Pred.
Name	Symbol	(km)	Pc	F	Pc	F	Pc	F	Mag.
Dengke	(S1)	80	0.02	0.03	0.04	0.05	0.03	0.04	7.6
Manigange	(S2)	150	0.17	0.72	0.14	0.69	0.16	0.71	7.8
Ganzi	(S3)	65	0.13	0.28	0.13	0.29	0.13	0.29	7.5
Zhuwo	(S4)	25	0.00	0.00	0.00	0.00			6.7
Luhuo	(S5)	80	0.00	0.00	0.00	0.00	0.00	0.00	7.5
Qiajiao	(S6)	50	0.15	0.17	0.16	0.20	0.16	0.19	7.3
Daofu	(S7)	35	0.05	0.05	0.06	0.06	0.06	0.06	7.1
Qianning	(S8)	50	0.30	0.44	0.28	0.45	0.29	0.45	7.3
Yalahe	(S9)	40	0.13	0.48	0.13	0.48	0.13	0.48	7.2
Selaha	(S10)	40	0.28	0.80	0.25	0.78	0.27	0.79	7.2
Kangding	(S11)	45	0.00	0.00	0.00	0.00	0.00	0.00	7.1
Moxi	(S13)	80	0.12	0.33	0.12	0.33	0.12	0.34	7.5
Shimian	(S14)	80			0.10	0.63	0.10	0.63	7.5
Lugu	(S15)	80	0.37	0.41			0.37	0.41	7.5
Puge	(S16)	90	0.00	0.00	0.00	0.00	0.00	0.00	7.6

Figure 7 illustrates the calculated probabilities. It can be seen from Figure 7 that 6 segments have relatively high cumulative probabilities that are equal to or higher than 0.45 till A. D. 2026. These 6 segments are all located at those seismic gaps identified from the space-time pattern of the historic ruptures (see Fig. 5 and Fig.7). However, not all these 6 segments have relatively high conditional probabilities   in the next 30 years. Actually, if a segment has a longer recurrence time, such as more than 300 yeas, to the next event, in a shorter time interval, such as D~30-years, the calculated conditional probability could be not h-h, no. matter the elapsed time from the last events is either longer or shorter. This is somewhat different from the case on inter-plate tectonic environment where the activity of inter-plate fault is so high that average recurrence time between large or great earthquakes is usually decades or between 100 and 200 years[9].

On the intra-plate tectonic environment of the China's mainland, it would be better to take not only conditional probability, but take cumulative probability to analyze long-term earthquake potential on fault segments. For example, segment 2 and segment 6 have the same high conditional probability (P_c=0.16) in the coming 30 years, but segment 2 has much higher cumulative probability (F=0.71, till A. D. 2026) than segment 6 (F=0.19, till A. D 2026). So segment 2 has higher earthquake potential than segment 6 in the coming 30 years.

From the comparisons among or between probabilities of different fault segments, and from the light of long-term forecast, the author suggests that two areas, from Qianning to KangdinS and from Shimian to Xichang (see Fig. 6 and Fig. 7), should be considered as the main risk areas for the coming 30 years. The former contains fault-segments 8 to 11, and the latter contains segments 14 and 15.

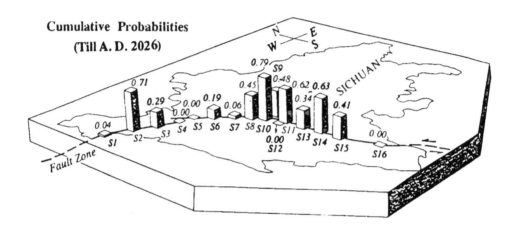

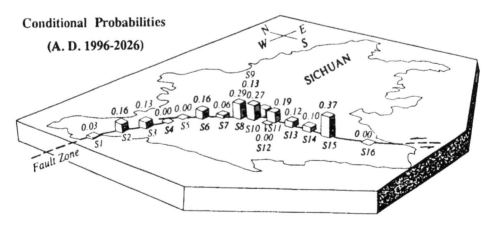

Fig. 7. Diagrams showing calculated probabilities that represent future earthquake potential on the fault segments The fine-curve on the surface is the boundary of Sichuan province, and the thick curve represents the studied fault zone. Column heights are proportional to probabilities.

DISCUSSION

This paper has carried out seismic potential evaluation for the main active strike-slip fault zone in western Sichuan, China. The follows should be emphasized: (1) This work is only a preliminary effort. It is obvious that the result exits uncertainties. This is mainly caused by geologic data uncertainties.These data include fault slip-rates, mean coseismic slips, paleoseismic dating, and elapsed times for several segments. (2) The result's uncertainties are also caused by the model uncertainty. The generic probabilistic distribution of recurrence time for characteristic earthquakes[8] is setup on the data from inter-plate earthquakes. Whether this distribution could be used on the intra-plate tectonic environment like the China's mainland is still a question. In the case of having no any selection else, the result of employing this model is an approximation only. (3) No matter how rough the result of thi9 study is, to a certain degree, it is still useful to the long-term seismic hazard evaluation for the studied region. The fact, those fault segments with relatively high cumulative probabilities indicate the seismic gaps of having no earthquake for long time, suggests that although we are not able to get accurate probabilities of earthquake occurrence by employing uncertain data and models, we can get at least the information about which segment having higher or lower seismic potential relative to other segments.

REFERENCES

1. C. R. Allen and others, Field study of a highly active fault zone: The Xianshuihe fault of southwestern China, *Geol. Soc. Am. Bull.*, 103, 1178-1199. (1991)
2. M. G. Bonilla, and others, Statistical relations among earthquake magnitude, surface rupture length, and surface fault displacement, *Seis. Soc. Am. Bull.*, 74, 2379-2411.(1984)
3. C. Cai and others, *Character of Late-Quaternary activity and assessment of seismic risk on the Ganzi-Yushu fault zone*, Report of the seismological sciences coordinated fund, Project No. 92091, State Seismological Bureau, 47pp(in Chinese).(1994)
4. China-Japan Symposium on Earthquake Prediction, 119-137, Seismological Press, Beijing.(1994)
5. Deng Q., and others, Relationships between earthquake magnitude and parameters of surface ruptures associated with historical earthquakes, *Research on Active Fault*, 2, 247-264, Seismological Press(in Chinese).(1992)
6. Li T. and others, The activity of the Selaha-Kangding-Moxi fault since the late Pleistocene(in Chinese), *Research on Active Fault*, 2, 1-14.(1992)
7. Li T. and others, Recent activity of the Zhedotang fault and the 1955 earthquake of magnitude 7.5(in Chinese), *Research on Active Fault*, 2, 15-23.(1992)
8. S. P. Nishenko, and R. Buland, A generic recurrence interval distribution for earthquake forecasting, *Bull. Seism. Soc. Am.*,77, 1382-1399.(1987)
9. S. P. Nishenko, Circum-Pacific seismic potential: 1989-1999, *PAGEOPH*, 135, 2, 169-259.(1991)
10. Qian H., and others, Prehistoric earthquake on the north segment of Anninghe fault and their implications for seismological research, *Earthquake Research in China*, 7, 4, 330-341.(1993)
11. K. Shimazaki, and T. Nakata, Time-predictable recurrence model for large earthquakes, *Geophys. Res. Lett.*, 7, 279-282. (1980)
12. D.B. Slemmons, Determination of design earthquake magnitudes for microzonation, in: *3rd International Earthquake Microzonation Conference Proceedings*, 199-130.(1982)
13. Wen X. and others, Neotectonic feature and earthquake risk assessment of the Ganzi- Yushu fault zone (in Chinese), *Seismology and Geology*, 7, 3, 23-32.(!985)
14. Wen X. and others, Recent slip rates, earthquake recurrence intervals and strong seismic hazards on the northwestern segment of the Xianshuihe fault zone, *Earthquake Research in China*, 2, 4, 432-451.(1988)
15. Wen X., Conditional probabilities for the recurrence of earthquakes on the Xianshuihe fault zone within the coming three decades (in Chinese), *Earthquake Research in China*, 6, 4, 8-16.(1990)
16. Wen X., Problem on quantitative study of active fault and probabilistic estimation of long-term seismic potential (in Chinese), *Research on Active Fault*, 1, 174-183, Seismological Press.(1991)
17. Wen X., Quasi-time-predictable recurrence behavior and probabilistic assessment of seismic potential on fault segments (in Chinese), *Earthquake Research in China*, 9, 4, 289-300.(1993)

18. Wen X., *Quantitative Estimates of Seismic Potential on Active Faults*, Seismological Press, 150 pp. (in Chinese). (1995)
19. Working Group on California Earthquake Probabilities, Probabilities of large earthquakes occurring in California on the San Andreas fault, *U. S. Geol. Surv. Open-File Rep.* 88-398.(1988)
20. Working Group on California Earthquake Probabilities, Probabilities of large earthquakes in the San Francisco Bay Region, California, *U. S. Geol. Surv. Circ.* 10~3, 51 pp. (1990)
21. Working Group on California Earthquake Probabilities, Seismic hazards in southern California: Probable earthquakes, 1994-2024, *Bull. Seis. Soc. Am.*, 85, 379-439.(1995)
22. Zhao G. and others, The late Quaternary slip rate and segmentation of the Xianshuihe active fault zone, In: *Proceedings of the PRC- USA Bilateral Symposium on the Xianshuihe fault zone*, 41-57, Seismological Press.(1992)

Proc. 30ᵗʰ Int'l. Geol. Congr., Vol. 5 pp. 115-121
Ye Hong (Ed)
© VSP 1997

Karakorum Fault and its Geodynamics

ZHENG JIANDONG

Institute of Geology, State Seismological Bureau 100029, Beijing, China

Abstract

Karakorum Fault is a major strike-slip fault lying in the central part of Asian Continent. Its northern part lies in Afghanistan, the central part is between Xinjiang and Kashmir, the southern part is in Tibet. It is obvious that the Karakorum Fault is an active fault. A series of stronger earthquakes occured along the northern part of the fault zone. The Karakorum Fault and Altun Tagh Fault are the two most prominent strike-slip faults in this region. The former is right lateral strike-slip fault, the later is left lateral strike-slip fault. Karakorum Fault creates a huge escaping tectonics in the eastern side of the Pamir Plateau. The rate of eastward tectonic escape is 5-10 mm/yr. Chaman Fault and Herat Fault also create a huge escaping tectonics in the western side of the Pamir Plateau. From the viewpoint of geodynamics, the Tibetan and Persian Plateaus are symmetrical to the Pamir collision corner and their eastward and westward tectonic escape rate are about 5-10 mm/yr respectively. In addition, there are a lot of intermediate earthquakes occurred in the West Kunlun and Karakorum Mountains. An intermediate earthquake of M_b=6.0 occurred at Yecheng in February, 1980 in the Tarim Basin. It made the seismic intensity of VII in the surface about 90 km north of the microscopic epicenter. This suggests that the southwest block of the Tarim Basin subduct southward beneath the West Kunlun and Karakorum Mountains.

Keywords: Karakorum Fault, Seismic deformation belt, Subduction, Geodynamics

INTRODUCTION

The West Kunlun and Karakorum Mountains lie in the western margin of China. Because it is located at high altitudes and in nonpopulated area, only a few geoscientists reached there and the geological research is sparse. From 1987 to 1992, scientists from Chinese Academy of Sciences (CAS) and Center National Researcher Scientique (CNRS) of France conducted " The Integrated Science Expedition on the Karakorum and Kunlun Mountains" [11]. Pan Yusheng and Wang Yi [3,4] studied tectonics of the Karekorum and West Kunlun Mountais. Jiang Chunfa and others [1] studied geological evolution of the whole Kunlun Mountains. Qin Guoqing et al.[4] conducted MT in the region. In 1987, the author of this paper conducted paleomagnetic study along West Kunlun Mountains [8]. Later-on, in 1992 the author participated in the International Symposium on the Karakorum and Kunlun Mountains in Xinjiang and made a field investigation in the region from Kashi to Kunjirap. This paper briefly discuss the features and geodynamics of the Karakorum fault.

GEOLOGICAL BACKGROUND

The Karakorum Mountains lies between Tibetan and Pamir Plateaus. The Karakorum Fault passes through the northern part of West Kunlun Mountains and Karakorum Mountains. Tectonically, the region can be subdivided into three tectonic zones, i.e. the north Kunlun tectonic zone, the central Kunlun tectonic zone, and the Karakorum tectonic zone (Fig. 1).

Zheng Jiandong

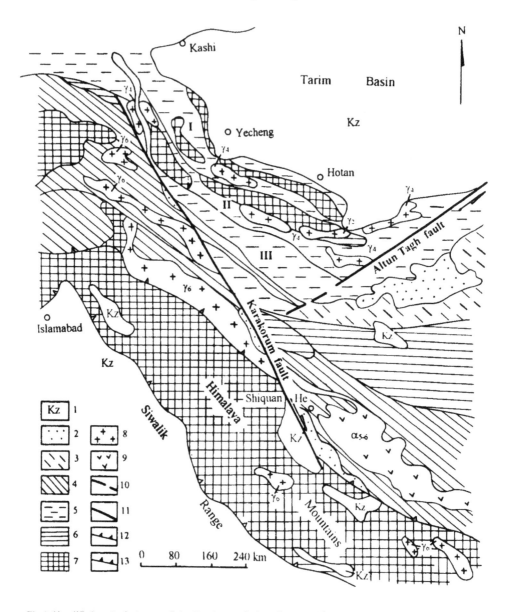

Fig.1 Simplified geological map of the Karakorum fault and surrounding areas. 1. Cenozoic; 2. Jurassic; 3. Triassic; 4 .Mesozoic; 5. Upper Paleozoic; 6. Lower Paleozoic; 7. Precambrian; 8. Granite; 9. Volcanic rock; 10. Fault; 11. Strike-slip fault; 12. Thrust fault; 13. Suture; I. North Kunlun tectonic zone; II.Central Kunlun tectonic zone; III. Karakorum tectonic zone.

The central Kunlun tectonic zone is a principal tectonic unit. It controls the geological evolution of the region. It consists of Precambrian rocks, Paleozoic and Mesozoic sedimentary rocks and poly magmatic rocks. A series of intrusive rocks, such as granite and diorite are developed in the central Kunlun tectonic zone. The ages of these rocks are mainly 540-400 Ma and 260-200 Ma. These rocks belong to calcalkine rock series. Pan Yusheng [2] suggested that there are double arc-island belts in the Caledonian and Indosinian stages respectively, and that the Mazar-Kangxiwar Suture lies northwards. The north Kunlun tectonic zone also consists of Paleozoic and Mesozoic rocks, whereas the Precambrian rocks are limited and the igneous rocks are less than the central part. Therefore, it would belong to passive continental margin in the Paleozoic. The Karakorum tectonic zone lies south of the central Kunlun tectonic zone. Its southern boundary is the Karakorum Fault. It mainly consists of Paleozoic and Mesozoic rocks characterized by Gondwana formations and glacial animals. This provides the scope and boundary of the Gondwana Land in the geological history. On the other hand, in the Hotan and Mingfeng areas, a series of reverse faults and nappes thrust northwards. It is observed that some parts of the north Kunlun tectonic zone thrust to the Tarim Basin. There is no obvious boundary between Kunlun tectonic zone and Tarim Basin.

FEATURES OF THE KARAKORUM FAULT

The Karakorum Fault, a major NW trending right lateral strike-slip fault belt in the central Asia, extends from Mingteke Daban,   Kunjirap, Qogir Peak,   Ritu to Shiquan He where it joins the Gar He fault. The total length of this fault is about 1000 km. Its northern part lies in Afghanistan, the central part is between Xinjiang Uygur Autonomous Region and Kashmir, and the southern part is in Xizang Autonomous Region (Tibet). It is a remarkable fault. The fault offsets the ophiolites by about 250 km between the Yarlung Zangbo River and Indus River. It also controls the distribution of the Jurassic formation and the Mesozoic -Cenozoic volcanic rocks in the northern slope of the Gangdise Mountains. The Karakorum Fault zone is very clear on Landsat image. The fault zone consists of a series of en echelon faults. It links the Gar He Fault in southern Tibet.

It is important to discuss the relationship among Karakorum Fault and other faults. P.Tapponnier [5] showed that the Altun Tagh Fault curved in the Aksayqin Lake and it is linked with the Mazar-Kongxiwar Fault. However our observation does not support this interpretation. Our data suggest that the Altun Tagh Fault is a straight left lateral strike-slip fault, which does not mix up with other faults [9,10]. The Altun Tagh fault cut the Mazar-Kongxiwar Fault and does not cross the Karakorum Fault. In addition, the Karakorum Fault does not link with the Fergana Fault. The former ends in the Pamir corner and the later extends to the western margin of the Tarim Basin. These two fault zones are also characterized with en echelon style trending NW direction.

There are a series of rifted basins along the Taxkorgan valley, such as the Muji Basin, Baoziya Basin, Tagarma Basin, Taxkorgan Basin, forming a string of basins . A lot of active faults extend along both sides of these basins. We named these active faults the Taxkorgan Fault. These active faults are mainly normal faults. Some of them are associated with seismic deformation belts (earthquake surface ruptures). Based on the field investigations, we identified three seismic deformation belts in the Taxkorgan valley (Fig.2).

(1) Seismic deformation belt along western margin of Kongur Mountain
The northern and southern parts of this belt pass through the eastern margin of the Muji and Baoziya Basins in NW direction, whereas the central segment extends near Bulunkou in N-S direction. A three meters high fault scarp occurs at Bulunkou. It offsets the gullies and ridges. The earthquakes of M= 5.0 and M= 6.4 occurred near the Kongur Mountain on 14 July, 1953 and 15 November, 1959.

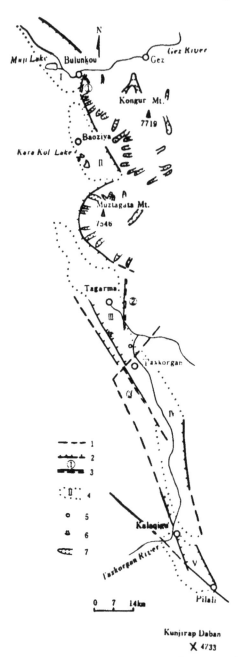

Fig.2 Distribution of the Taxkorgan fault and seismic deformation belt. 1. Active fault; 2. Normal fault; 3. Seismic deformation belt; 4. Rifted basin; 5. Sandblow; 6. Hot spring; 7. Recent glacier, I. Muji Basin; II. Baoziya Basin; III. Tagarmation Basin; IV. Taxkorgan Basin; V. Karaqingu Basin; (1) Seismic defomation belt on the western margin of Kongur Mt; (2) Seismic defomation belt on the eastern margin of Tagarma Basin; (3) Seismic deformation belt on the western margin of Taxkorgan basin.

(2) Seismic deformation belt along eastern margin of Tagarma Basin

It extends from north to south between glaciofluvial fans and mountain roots along eastern margin of Tagarma Basin. The total length of this fault is about 40 km. The width is about 20-80m. The deformation belt is characterized by normal fault with landslides, tensile cracks, grabens and horsts and earthquake ruptures in the surface. The earthquake of M=7 on 5 July, 1895 occurred here and formed the seismic deformation belt. A wonderful sandblow is found about 20 km south of the seismic deformation belt. Another two active faults are found along western margin of the Tagarma Basin. The east branch passes through the Tagarma hot spring.

(3) Seismic deformation belt along western margin of the Taxkorgan Basin

It extends from NNW to SSE about 30 km long. A series of landslides are distributed between glaciofluvial fans and mountain roots along a line. In history many strong earthquakes occurred in the Taxkorgan area (Table 1).

GEODYNAMICS

The Karakorum and Altun Tagh Faults are two major strike-slip faults. They form a conjugate shear couple in the northwestern margin of the Tibetan Plateau . The former is a right lateral strike-slip fault trending NW, the later is a left lateral strike-slip fault trending NEE. These two faults create a huge escaping tectonics in the eastern side of the Pamir collision corner, which makes the Qiangtang block, the northern part of the Tibetan Plateau moving eastwards (Fig.3). The rate of eastward tectonic escape is 5-10 mm/yr [7]. On the other hand, the

Chaman and Herat Faults also create a huge escaping tectonics in the western side of the Pamir collision corner. Thus, the Tibetan Plateau and the Persian Plateau are symmetrical to the Pamir collision corner and their eastward and westward tectonic escape rate are about 5-10 mm/yr respectively.

Table 1. Earthquakes of M≥5 in Taxkorgan area

Data	Latitude	Epicenter Longitude	location Epicenter	Magnitude	Depth
	(North)	(East)	(Area)		(km)
1895-07-05	37.7°	75.1°	Taxkorgan	(7)	
1953-05-29	38.2°	75.1°	Taxkorgan	5	
1953-07-14	38.5°	75.1°	Kongur Mt.	5	160
1953-12-21	38.0°	75.1°	Taxkorgan	5 1/4	100
1959-11-15	38.7°	75.3°	Kongur Mt.	6.4	40
1972-01-13	37.7°	75.0°	South of Taxkorgan	5.4	84

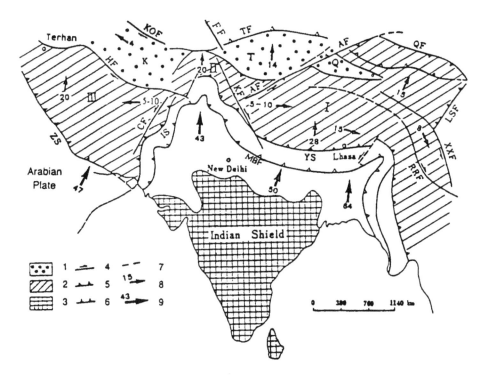

Fig. 3 Schematic geodynamics of the western part of Asian Continent. 1. Cenozoic continental basin; 2. Young plateau; 3. Precambrian shield; 4. Strike-slip fault; 5. Thrust fault; 6. Collision zone; 7. Uncertain fault; 8. Rate and direction of movement of the plate(mm/yr); 9. Relative rate and direction of movement of the plate (mm/yr); I. Tibetan Plateau; II. Pamir Plateau; III. Persian Plateau; K. Karakum Basin; Q. Qaidam Basin; T. Tarim Basin; AF. Altun Tagh fault; CF. Chaman Fault; F. Fergana Fault; HF. Heart fault; IS. Indus suture; KF. Karakorum fault; KOF. Kopet fault; LSF. Longmenshan fault; MBF. Main boundary fault; QF. Qilian fault; RRF. Red River fault; TF. Tian Shan fault; XXF. Xianshuhe fault; YS. Yarlung Zangbo suture; ZS. Zagros suture.

Seismologically, there are a lot of intermediate earthquakes occurred in the eastern side of the Pamir Plateau. Fig.4 shows distribution of the seismic focus (M≥3) in the western periphery of the Tarim Basin from 1965 to 1984. Wang Suyun et al.[6] carefully studied the earthquake

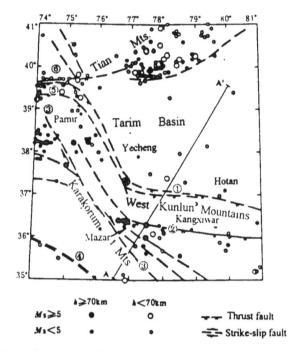

Fig. 4 Distribution of seismic focus in Yecheng, Xinjing and surrounding areas (After Wang, et al. 1992). 1. Thrust fault; 2. Strike-slip fault; (1) North boundary fault of West Kunlun Mts; (2) Kangxiwar fault; (3) Karakorum fault; (4) Indus suture; (5) Pamir fault; (6) South boundary fault of tian Mts; ⊗ Macroscopic epicenter of Yecheng earthquake occurred on 14 February, 1980: ■ Micrososcopic epicenter of Yecheng earthquake occurred on 14 February, 1980.

of Mb=6.0 occurred at 36.40°N, 76.90°E, in Yecheng City of the Tarim Basin on 14 February, 1980 and pointed out that there are two seismic zones along the southwestern margin of the Tarim Basin. One is an intermediate seismic zone which reaches 100 km in depth. The other is a shallow seismic zone located below the Karakorum Mountains south of the intermediate seismic zone. The Yecheng earthquake occur within the intermediate

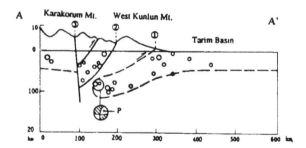

Fig. 5 Seismotectonic profile along the southwestern margin of Tarim Basin. (1) North boundary fault of the West Kunlun Mts; (2) Kangxiwar fault; (3) Karakoum fault; P. Paxis shwing shwing earthquake mechanism of the Yecheng earthquake.

seismic zone (Fig.5). Regional seismotectonic data show that the macroscopic epicenter (epicenter determined by isoseismal) located at the upper termination of the intermediate seismic zone, whereas the microscopic epicenter (epicenter determined by seismograph) located at the surface projection of the hypocenter. So the strong seismic intensity VII in the surface would be located 90 km north of the microscopic epicenter (37.30 °N, 76.90°E). This phenomenon implies that the southwestern block of the Tarim Basin subducts southwards beneath the northwestern margin of the West Kunlun and Karakorum Mountains. This explains why the neotectonics and seismicity is so strong in this region.

Acknowledgments

This study is supported by the "Item of Recent Crustal Movement and Geodynamics of China", which belongs to the Scientific Climb Program during the Eighth Five-Year Plan period (1991-1995). The author would like to express best thanks to the Seismological Bureau of Xinjiang Uygur Autonomous Region for their kind help in the field investigations. Also thank Professor Ye Hong for reading and improving the early manuscript of this paper and my wife Fan Heping for typewriting the paper.

REFERENCES

1. C.F.Jiang. *Opening-closing tectonics of Kunlun Mountains*: Geological Publishing House, Beijing (1992).
2. Y.S.Pan. Tectonic features and evolution of the West Kunlun region: *Scientia Geologica Sinica*, 4, 232-241 (1990).
3. Y.S.Pan and Y.Wang. Geological characteristics and tectonic evolution of the Kunlun and Karakorum ranges, *Advances in Geoscience*, 2, 9-21 (1992).
4. G.Q.Qin, J.H.Chen, D.J.Liu, Q.Gu and Y.W.Xiong. The characteristics of the electrical structure of the crust and upper in the region of the Kunlun and Karakorum Mountains: *Acta Geophysica Sinica*, 37:2, 195-199 (1994).
5. P.Tapponnier. Subduction, crustal folding and slip partitioning along the edge of Tibet: International Symposium on the Karakorum and Kunlun Mountains, *Abstract*,China Meteorological Press, Beijing, p.8 (1992).
6. S.Y.Wang, Z.L.Shi, and W.L.Huan. The Yecheng, Xinjiang earthquake of February 14, 1980 - A destructive intermediate-focus earthquake: *Acta Seismologica Sinica*, 14:2, 137-143 (1992).
7. J.D.Zheng. A preliminary study on geodynamics of Qinghai-Xizang (Tibet) Plateau: *Geoscience Journal* , Beijing Graduate School, Wuhan College of Geology, 4:2, 194-206 (1988).
8. J.D.Zheng, G.L.Cheng,Y.H.Bai, C.P.Zhou, Y.H.Sun, Q.G.Sun and S.L.Li. Geological significance of paleomagnetic testing for the Meso-Cenozoic Era in the Tarim and Qaidam Basin: *Earthquake Research in China*, 3:4, 447-455 (1989).
9. J.D.Zheng. Significance of the Altun Tagh fault of China: Episodes, 14:4, 307-312 (1991).
10. J.D.Zheng. Karakorum fault and Taxkorgan seismic deformation zone: *Seismology and Geology*, 15:2, 107-116 (1993).
11. D.Zheng, Q.S.Zhang and Y.S.Pan. Preface in: *Proceedings of International Symposium on the Karakorum and Kunlun Mountains*: China Meteorological Press, Beijing (1994).

Proc. 30 *Int'l. Geol. Congr., Vol. 5 pp. 123-131*
Ye Hong (Ed)
© VSP 1997

Summary of Historic Faults in Taiwan

MING-SHENG YU[1], SHIH-NAN CHENG[2] AND YEON-TIEN YEH[2]

1). *Central Geological Survey. P. O. Box 968, Taipei, Taiwan*
2). *Institute of Earth Sciences, Academia Sinica, P.O. Box. 1-55, Nankang. Taipei, Taiwan*

Abstract

In order to figure out the relationships between historical earthquakes and the associated faults, their seismological data are restudied. A Monte Carlo algorithm is used to reconstruct earthquake source parameters based on P-S times for instrumental historical event where the timing systems of seismic stations available were not in that time. Comparing the isoseismal map and the trajectory of surface faulting with relocated epicenters. it can be found that results of this study may be more reasoning than the previous studies. The following points are noted: 1. four of the ruptures occurred in the coastal plain and foothills of western Taiwan and four occurred in the Longitudinal Valley of eastern Taiwan; 2. there are several features suggest that the Chihshang fault and the Chimei fault were activated during the movement of the earthquakes but has not been reported before; 3. the Shinchoshan fault and the Juisui fault which associated with 1935 and 1972 earthquakes were no sign of surface faulting. they are reinterpreted just landslide induced by earthquakes. According to the stress analysis of those earthquake-faults, the steadiness of NW-SE principal stress of compression is witnessed.

Key words: Historic Fault. Monte Carlo Algorithm, Relocation, Fault Plane Solution, Taiwan

INTRODUCTION

Active faults that have produced surface breaks during major earthquakes in historic times are know in the coastal plain and hilly areas of western Taiwan and in Hualien and the Taitung Longitudinal Valley of eastern Taiwan but none have been discovered, either by surface exploration or geophysical means, in the Central Range [2,6,19]. There are some differences in the attitudes of earthquake faults between eastern and western Taiwan. In eastern Taiwan, left lateral slip and NNE strike predominate. But in western Taiwan. they exhibit mainly right lateral slip and NEE strike. Historical faults in Taiwan, especially those in the eastern part, are believed to be closely related to the regional tectonic history of western Pacific, especially the interaction between the Philippine Sea Plate and the Eurasian plate [19].

ANALYSIS

Instrumental observation of Taiwan earthquakes began in 1897 when the first seismograph was installed at the Taipei weather observatory. Later, many kinds of seismographs were installed at other weather observatories throughout Taiwan to form a network. Because lacks of independent time system in the period of 1898-1972, there are only P-S times to be use in locating process, the epicenter and focal depth was estimated from iso-P-S time graph. Although the error of result is relatively higher than the ones after 1973, this method was a better choice under the environment of data quality and seismograph in that time. the results of location are still practical. But for some large earthquakes that the deviation between the location and surface fault is too large to neglect for studies in seismicity, seismotectonics, earthquake prediction and seismic risk analysis, it is necessary to be re-estimated the historic faults. Under this condition, A Mote Carlo algorithm is applied to re-locate the instrumental historical earthquakes based on P-S times [9]. The results of inversion are combined with geological data (such as strike, dip, and dislocation of surface fault) and polarity of first P motion to estimate the possible fault-plane solution [3,9,12]. In this study, the parameters to be evaluated were the latitude, longitude and depth of earthquake focus. From the iso-seismal map and distribution of surface fault, there are some a priori information to limit the distribution range of hypocenter parameters. The observed data of P-S times [1,7, 8] and the crustal velocity model [10] are used for calculating the seismic travel times.

RESULTS AND DISCUSSIONS

Faulting associated with six large earthquakes in Taiwan from 1906 to 1972 has displaced the ground surface by amounts ranging from 1 to 3 m. Four of the ruptures occurred in the coastal plain and foothills of western Taiwan and four occurred in the Longitudinal Valley of eastern Taiwan. Some equivocal evidence suggests fault creep on the faults but no creep has been reported based on the measurements made since 1975. The fault ruptures are described in chronological order.

17 March 1906 Event
Surface faulting near Meishan in western Taiwan was associated with a magnitude 7.1 earthquake on 17 March 1906 that killed 1,258 people and destroyed more than 6,000 houses. The fault trends N 80°E over most of its length but toward the west it trends N 50°E and then extends nearly due west to Minhsiuang. The fault trace, which was incompletely mapped in 1906, extended about 13 km from Hsinkang to Meishan and Tahu, being a right-lateral slip fault. This fault, which has formed a scarplet, truncates the terrace deposits. Maximum displacement dextral strike-slip was 2.4 m at the point near the east end of the fault where 1.2 m of vertical displacement also occurred. At this point and southwest of it the north side of the fault was relatively lowed. but on the short segment northeast of this point the south side of the fault was relative lowed. The maximum vertical displacement of 2.1 m occurred near the northeast end, combined with a right slip of 1.8m. Part of the 1906 scarp was still visible in 1996 near Meishan but it has been artificially modified and is only about half as high as it was 1906 at that point [8].

The hypocenter was relocated at 23.585°N and 120.535°E with depth of 15 km. This event was related to a dextral strike-slip fault, striking N80°E. The average rake is about -144° if estimated from the surface of the of the Meishan area. Assuming the dip angle is vertical as the geophysicist's suggested, i.e., 90°, the fault-plane solution shows stress axis to be 131°/25°. Maximum compression axis of this shock is closely oriented NW-SE which agree with the direction of regional tectonic stress. [13]

21 April 1935 Events
Surface rupture occurred on two faults in northwestern Taiwan on 21 April 1935 accompanied by an earthquake that killed more that 3,000 people and destroyed more than 17,000 houses. The faults were called the Szetan and Tuntzuchio faults. respectively.

The Szetan fault, located east of Miaoli, originated during the earthquake on April 21, 1935. The surface rupture on the Szetan or northern fault was 15 km long, the northern 5 km of which consisted of short discontinuous ruptures. The fault dips 85°W and is of reverse type, with the west side uplifted 3 m relative to the east side, and had a small right-slip component in places.

Over most of its length the Szetan earthquake fault parallels but lies more than 0.5 km east of the Szetan fault; only about 1.5 km of its north end is approximately coincident with a mapped fault. Folds in the area consist of northeast-trending anticlines and synclines with axial reverse faults. The thrust nature of the Szetan rupture is consistent with the stress field that produced the Tuntzuchio fault and suggests that both ruptures originated from deep-seated tectonic stresses [2].

The southern rupture, called the Tuntzuchio fault, extended for approximately12 km, consisting of a series of en echelon breaks of various lengths with right slip of as much as 1.5 m and vertical displacement of as much as 1 m, with the northwest side generally downdropped except for about 1 km of its length at the northeast end, where the southeast side was relatively down. It is a right-lateral slip fault, striking N60°E for a total distance about 20 km. The maximum horizontal and vertical displacement are 200 cm and 60 cm respectively. It cuts through terrace deposits, and alluvium [2].

The hypocenter of the second shock relocated at 24.600°N and 120.900°E at 2 km in depth may be related with the rupture of Szetan fault, a thrust fault, striking N67°E, dipping 85°NE and raking 90°. Fault plane solution shows the stress axis to be 112°/8°. The first shock was relocated at 24.600°N and 120.900°E at 5 km in depth may be related with the rupture of Tuntzuchio fault, a dextral strike-slip fault, striking N23°E, dipping 50°NW and raking 180°. Fault plane solution shows the stress axis to be 112°/5°. Maximum compression axes of these two shocks are closely oriented NW-SE which agree with the direction of regional tectonic stress [13].

5 December 1946 Event

The 1946 faulting occurred in the coastal plain and foothills in the southwestern part of Taiwan about 10 km northeast of Tainan. The associated earthquake of magnitude 6.1 killed 74 persons and distroyed 3,577 houses. The fault showed consistent right-lateral slip, with the southeast side relatively down where it had a clearly defined trace. This type of fault has been designated as dextral concave-arc-transform fault. the maximum right slip of 2.0 m and maximum dip slip of 0.76 m occurred at the same place. The clearly defined surface rupture extended at least 6 km with a trend of N80°E but the total length of the fault in the subsurface can be traced for 12 km and dislocated the Holocene Tainan Formation. Groups of en echelon cracks and warps with small vertical displacements were found in a linear zone extending about 6 km southwest of the surface fault.

The hypocenter was located at 23.075°N and 120.325°E with depth of 5 km. This event was related to a dextral strike-slip fault, striking N 80°E. The average rake is about -152° if estimated from the slickensides of the outcrop. Assuming the dip angle is vertical as the geophysicist's suggested. i.e., 90°, the fault-plane solution shows stress axis to be 129°/20°. Maximum compression axis of this shock is closely oriented NW-SE which agree with the direction of regional tectonic stress [13].

22 October 1951 Event

The surface ruptures occurred on the Meiluan fault near Hualien on October 22 associated with an earthquake of magnitude 7.1 which killed 45 persons and destroyed 522 house. The length of faulting is given variously as 5 to 7 km which can he taken as minimum, inasmuch as the fault extended beneath the sea at the north end.

The first earthquake (05:34) on October 22, 1951, occurred east of Hualien, with accompanying surface rupture, and 6 hours later (11:29). another shock near north of Hualien. These surface   faults were close to the Meiluan fault. Maximum reported fault displacements were 2 m of lea slip and 1.2 m uplift on its southeastern side. Attitude of the fault was obtained at Minyili and the vertical displacement was measured at the north tip of the fault at Chihsingtan [5,14].

The hypocenter of the second shock (11:29) relocated at 24.075°N and 121.725°E at 1 km in depth may be related with the rupture of Meiluan fault, a sinistral strike-slip fault with thrust component, striking N 25°E, dipping 85°SE and raking 73°. Fault plane solution shows the stress axis to be 130°/38° [18].

The first shock (05:34) was relocated at 23.875°N and 121.725°E with depth of 4 km. This event was related to a thrust fault with sinistral strike-slip component. striking N25°E. The rake is about 31°if estimated from the surface offset of the Hualien area. Assuming with the same dip angle as the second shock (11:29), i.e., 85°SE, the fault-plane solution shows stress axis to be 154°/18°. Maximum compression axes of these two shocks are closely oriented NW-SE which agree with the direction of regional tectonic stress [16].

It is worth noting that the P-S times at the Taitung station of the first shock (05:34) (5.8 sec.) is too fast if it started from the Meiluan fault, northern Hualien and traveled through the whole Valley. In addition, a single Meiluan event should not be enough to explain the fractures observed from Changpin to Chengkuan. This phenomenon indicated that there did exist another shock, i.e., probably the Chihshang fault activated nearly at the same moment [17].

25 November 1951 Events
The surface rupture 1951 occurred on November 25 associated with an earthquake of magnitude 7.3 that caused 20 deaths, injured 326, and destroyed more than 1,000 homes. They are associated with a shock (02:47) east of Chihshang, and followed by another shock near Yuli 3 minutes later (02:50). Locations of these surface faults were closely to the Chihshang fault and the Yuli fault [11, 17]. The rupture was about 40 km long and the maximum reported displacement was 2.08 m of left reverse oblique slip. Subsidiary faulting may have occurred near the west edge of the Coastal Range about Tafu. It is reported the oblique movement with fresh slickensides in a scarplet near Tafu. The eastern side was upthrusted with a dip-slip of 131 cm and pushed northward with a strike-slip of 163 cm.

Using the method mentioned above, the hypocenter of (02:47) was relocated at 23.125°N and 121.225°E at a depth of16 km. The event may have created by a thrust fault (Chihshang fault) [15] with some sinistral strike-slip component, striking in N 32°E and dipping to 70°SE. The rake is 70°S. Fault plane solution shows the principal compression stress axis to be 137°/ 22°[3,11].

The shock (02:50) was relocated at 23.275°N and 121.350°E at a depth of 36 km. It may be associated with the rupture of the Yuli fault, a sinistral strike-slip fault with thrust component, striking N25°E. The rake angle (40°) was estimated from the slickenside of the surface offset. Assuming the dip angle to be the same as that of Chihshang fault, i.e., 70°SE, a fault plane solution 149°/10°was obtained. Again these two shocks are both NW-SE oriented and agree with the regional tectonic stress [3,11].

24 April 1972 Event

The third surface rupture in the Longitudinal Valley occurred on April 24. 1972 associated with an earthquake of magnitude 6.7 that caused 5 deaths, injured 17. and destroyed more than 100 homes. The approximate location of the surface faulting. which was closely in Fengpin [7,17]. The event was relocated at 23.638°N and 121.551°E with depth of 33 km and accompanied by a thrust fault striking N 44°E with some dextral strike-slip component. The fault plane solution showed the dip angle to be 60°SE and the rake 62°E. The axis of the principal compression stress of this solution is demonstrated to be horizontal and direction NWW-SEE (108°/7°) [4].

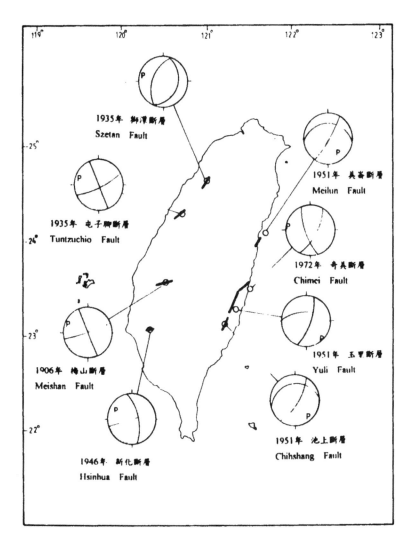

Fig 1. Fault plane solutions of the historic faults in Taiwan

A survey of the earthquake area had been taken by Central Weather Bureau after the shock [9]. Their observations indicated that the 'Juisui fault' which associated with 1972 earthquake were no sign of surface faulting. It was reinterpreted landslide induced by

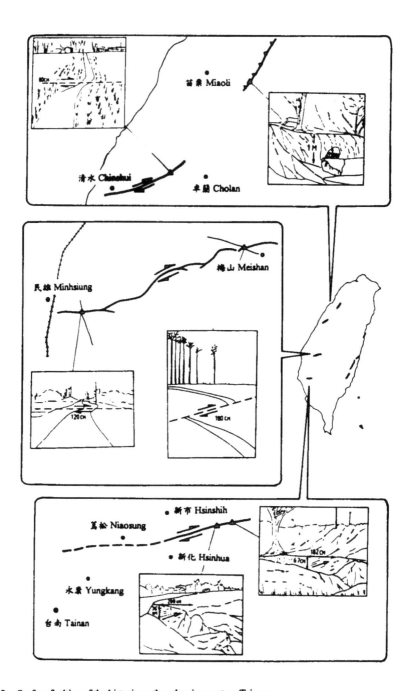

Fig.2   Surface faulting of the historic earthquakes in weastern Taiwan

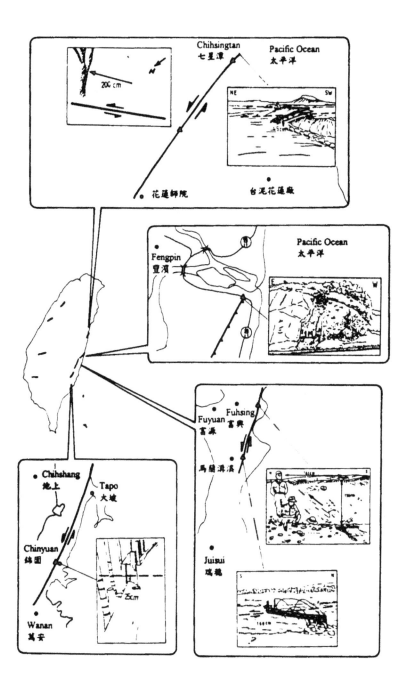

Fig.3 Surface faulting of the historic earthquakes in eastern Taiwan

M.S.Yu et al.

earthquake [17]. We have discovered some outcrops around the Chimei fault and the Yuli
fault were activated during the movement of the earthquake but has not been reported before.
It means the Juisui earthquakes may be associated with the rupture of dextral strike-slip
Chimei fault and sinistral strike-slip Yuli fault. [17].

The data on historic surface faulting are summarized in Table 1,2. and Figure 1,2,3.

Table 1. Maximum displacement of earthquake faults

Time	District	M	Fault	Length	Displacement	
					Vertical	Horizontal
190603170642	Chiayi	7.1	Meishan	13KM	210cm	240cm
193504210602	Taichung	7.1	Tuntzuchiao	20KM	60cm	200cm
193504210626	Taichung	5.8	Szetan	21KM	300cm	
194612050647	Tainan	6.1	Hsinhwa	6KM	76cm	200cm
195110221129	Hualien	7.1	Meiluan	1KM	120cm	200cm
195111250247	Taitung	6.5	Chihshang	40KM		
1195111250250	Taitung	7.3	Yuli	10KM	136cm	161cm
1197204241757	Juisui	6.7	Chimei	20KM		

Table 2. Fault plane solutions of the historic faults in Taiwan

Time	Epicenter	epth(km)	M	Fault	Fault plane	P-axis
190603170642	23-34.98 120-31.98	15.0	7.1	Meishan	N80°E/90°	131°/25°
193504210602	24-21.00 120-49.20	5.0	7.1	Tuntzchiao	N23°E/50°W	112°/5°
193504210626	24-42.00 120-54.00	2.0	5.8	Szetan	N67°E/85°W	112°/8°
194612050647	23-04.20 120-19.80	5.0	6.1	Hsinhwa	N80°E/90°	129°/20°
195110221129	24-04.50 121-43.50	1.0	7.1	Meiluan	N25°E/85°SE	130°/38°
195111250247	23-07.50 121-13.50	16.0	6.5	Chihshang	N32°E/70°SE	137°/22°
195111250250	23-16.50 121-21.00	36.0	7.3	Yuli	N25°E/70°SE	149°/10°
197204241757	23-30.71121-31.95	15.4	6.7	Chimei	N44°E/60°SE	108°/7°

Acknowledgments

The authors wish to express their gratitude to Prof. H.T. Chu of the Central Geological
Survey for the valuable suggestions during this study. They are also indebted to Miss C.L.
Hsiao of the Institute of Geophysics, National Central University for preparing the figures,
Prof. T.L. Hsu for providing some unpublished data of1951 Hualien and Taitung
earthquakes.

REFERENCES

1. Anonymous. *Report on 1951 earthquakes*, Central Weather Bureau, Taipei, Taiwan, 83 pp. (1952)
2. M.G. Bonilla. A review of recently active faults in Taiwan. U.S.G.S. *Open File Report* 75-41. Menlo Park,
   Calif., 58p. (1975)
3. S.N. Cheng, Y.T. Yeh and M.S. Yii: The 1951 Taitung earthquake in Taiwan: *J. Geol. Soc. China*, 39, 3,
   267- 285 . (1996)
4. S.T. Chiang, Y.B. Tsai and J.H. Wang. Relocation of main aftershocks of April 24, 1972 Juisui earthquake,
   Proc. Taiwan Symp. Geophys., 61-74. (1986)
5. T.L. Hsu. Recent faulting in the Longitudinal Valley of eastern Taiwan, *Mem. Geol. Soc. China*, 1, 95-102.
   (1962)
6. T.L. Hsu and H.C. Chang. Quaternary faulting in Taiwan. *Mem. Geol. Sec. China*, 3, 155-165.(1979)
7. S.M. Lu, Y.R. Hsu and N. Shih. *Report on Juisui earthquake of Apr. 24, 1972*, Central Weather Bureau,
   Taipei, Taiwan, 82pp. (1976)
8. F. Omori. Preliminary note on the Formosa earthquake of March 17, 1906: *Imp. Earthquake Inves. Comm.
   Bull.* 2, 53-69. (1907)
9. A. Tarantola. *Inverse problem theory, methods for data fitting and model parameter estimation*. Elsevier
   Science Publishing Company, New York, 613 pp. (1987)

10. Y.H. Yeh and Y.B. Tsai. Crustal structure of central Taiwan from inversion of P wave arrival times, *Bull. Ins. Earth Sci.* **1**, 83-102. (1981)

11. Y.T. Yeh, S.N. Cheng and M.S. Yu. A revisit on 1951 Taitung earthquake, *3rd R.O.C. and Japan joint seminar on natural hazards mitigation,* 74-88. (1993)

12. Y.T. Yeh, S.N. Cheng and M.S. YU. Relocation of several historical earthquakes and their fault-plane solution in Taiwan : *1995 Annual meeting extended abstract, Geol. Soc. China,* 469-473. (1995)

13. M.S. Yu and S.N. Cheng. On the earthquake-faults of western Taiwan : Proc. The Joint Symp. on Taiwan Quaternary [5] and on Investigation of Subsurface Geology1*Engineering Environment of Taipei Basin,* National Central University, Chunli, 1-7. (1994)

14. J.E. York. Quaternary faulting in eastern Taiwan, *Bull. Geol. Surv. Taiwan,* **25**, 63-75. (1976)

15. M.S. Yu, H.T, Chu, C.S. Hou and J.C. Lee. The Chengkung Earthquake of May 29, 1992 and the Chihshang Fault: *Bull. Central Geol. Sur.,* **9**, 107-121. (1994)

16. M.S. Yu. Wrench-Fault Characteristics of the Taitung Longitudinal Valley Fault Zone : *Ti-Chi - 14,* **1**, 121-147. (1994)

17. M.S. Yu and S.N. Cheng. On the earthquake-faults of eastern Taiwan : *1995 Annual meeting extended abstract, Geol. Soc. China.* 464-468. (1995)

18. M.S. Yu. The Meiluan earthquake-fault as a miniature transform fault : *Symposium on Taiwan strong motion instrumentation program* (11), 160-165. (1996)

19. M.S. Yu. The earthquake-faults of Taiwan : *Proc. 2nd. Across-the-Straits Confer. On Earthquake,* 250-255. (1996)

Proc. 30ᵗʰ Int'l. Geol. Congr., Vol. 5 pp. 133-139
Ye Hong (Ed)
© VSP 1997

1995 Hyogoken-Nanbu (Kobe) Earthquake and Urban Active Faults

TAKAO MIYATA[1], JINGPENG HONG[2], YASUHIDE NIGAURI AND YASUO MAEDA[3]

1)*Department of Earth and Planetary Sciences, Faculty of Science, Kobe University, Nada-ku. Kobe 657, Japan.*
2)*Graduate School of Science and Technology, Kobe University, Nada-ku, Kobe 657, Japan.*
3)*Institute for Science and Mathematics, Education Development, University of Philippines, Diliman, Quezon1101, Philippines.*

Abstract

We derived the co-seismic ground displacement from the inclination of utility poles and the slippage of house inlet in the urban area of Kobe during the 1995 Hyogoken-Nanbu Earthquake. Linear distributions of large displacement appeared along the concealed active faults beneath Kobe. Also the concentrated zones of breakage appeared along the concealed active faults. Judging from these features and analysis of strong ground motion seismograms during the earthquake, we believe that the remarkable ground displacement indicative of the linear distributions was caused by faulting on the concealed active faults.

Keywords: Kobe Earthquake, urban active faults, co-seismic ground displacement

INTRODUCTION

The disaster earthquake with magnitude 7.2 (JMA) hit big cities in the Kobe-Osaka area, southwestern Japan, on January 17,1995. As a result of the fault movement on the seismogenic fault beneath Kobe, the earthquake took more 6,400 lives, and destroyed too many houses. reinforced concrete (RC) structures and lifelines.

After the earthquake, some concealed faults beneath the Kobe-Osaka area, Awajishima Island and Osaka Bay became clear, based on seismic reflection surveys of the Hyogo Prefecture [5], the Geological Survey of Japan [2] and the Committee of Earthquake Observation and Research in Kansai Area (CEORKA) [10].

The purpose of our study is to discuss the movement of concealed active faults beneath Kobe, based on co-seismic displacement derived from inclined utility poles and deformed house inlets. and co-seismic surface breakage.

SEISMOLOGY AND GEOLOGIC SETTING

The 1995 Hyogoken-Nanbu (Kobe) Earthquake (M_{JMA}7.2) occurred within the Rokko-Awaji Fault Zone. The mainshock was located at Akashi strait between Kobe and Awajishima Island (Fig. 1). The focal mechanisms of mainshock is under the east-west directional compression by the subduction of the Pacific plate. The aftershock during the first 10 hours after the mainshock [13] indicates the linear distribution of 25 km long to the northeast and 20 km long to the southwest from the epicenter. Kikuchi [7] and Irikura [6] pointed out respectively that the earthquake was divided into three sub-events corresponded with three fault segments, based on the analysis of teleseismic: body-wave and strong ground motion seismograms. The third sub-event is located in the Kobe area.

Some northeast-striking active fault run from Rokko Mountain to Awjishima Island [4]. Although the earthquake fault, called the Nojima Fault, appeared in the island [11], it did not become visible on the urban surface of Kobe. However, the large ground moves occurred and too many breakage was formed on the urban surface during the earthquake.

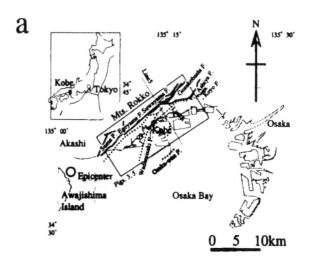

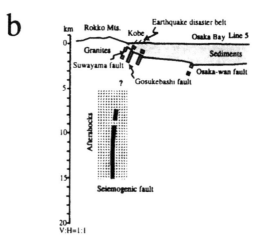

Fig. 1 a) Index map b) Schematic cross section from the Rokko mountains to the Osaka Bay (modified from [5]). Solid lines; Faults with the activity during the 1995 Hyogoken-Nanbu Earthquake. Dotted lines; Faults without the activity.

The geology of the Rokko Mountains is mainly Cretaceous granites, while the Kobe urban area is composed chiefly of Plio-Pleistocene sediments, having the thickness of about 1500 m [5]. Active faults are developed between the Rokko Mountains and the Osaka Basin. Some of them are buried under the sediments (Fig. 1). The topography of the Kobe urban area is classified mainly into fans, old rivers-valley floors, levees, raised bed rivers, back mashes, deltas, bars, Alluvial lowland and filled-up lands [15].

GROUND DISPLACEMENT

Co-seismic inclination of utility poles
A lot of utility poles in the urban area of Kobe inclined and broke during the earthquake. The utility pole moved due to seismic force (Fig. 2), and an opening, S, was produced between the inclined utility pole and asphalt surface on a roadway. We measured the S of 1307 utility poles (slim type), having the lengths of 15 m and 16 m, in the urban area during three months after the earthquake. The maximum value of the S was 8 cm long.

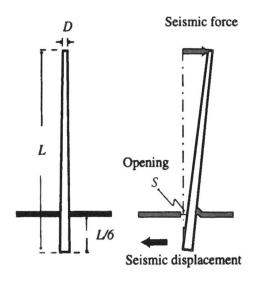

Fig. 2 Inclination of a utility pole of the slim type due to seismic ground moves. L: Length of the pole, d: Diameter of the top of the pole, F: Seismic force, S: Opening amount between the pole and an asphalt layer of roadway.

Figure 3 shows a distribution of the opening amount, S, between a utility pole and asphalt surface on a roadway. The eastern Kobe has lager value of the S than the western Kobe. Large amounts (2 cm or over) of the S indicate spots and linear distributions. Some linear distributions are located along the southwest extension (concealed fault) of the pre-existing active faults (Gosukebashi Fault, Ashiya Fault, Koyo Fault etc.) of the Rokko Mountains. While the spots were situated on soft, ground such as old ponds, old rivers, valley floors, back marshes and margins of fluvial fan.

Co-seismic slippage of house inlets
The horizontal slippage of a house inlet, having an inside diameter of 35 cm, was commonly produced at its joint, ranging 40-70 cm deep, during the earthquake. The co-seismic

displacement of the 1277 house inlets in the urban area of Kobe was investigated in order to determine the orientation of ground displacement, and the magnitude of impulsive force. Figure 4 shows a model of the displacement of a house inlet. due to impulsive force during the earthquake [14]. According to this model, the impulsive force is related to the depth, d, and the displacement amount, s. We can estimate the impulsive force, Δp, using d $\sqrt{s}$ [cm$^{3/2}$] for all displaced house inlets in the urban area of Kobe.

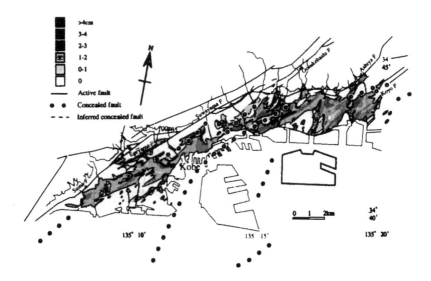

Fig. 3 Distribution of displacement of utility poles. Large amount of 2 cm or more indicates spots and linear distributions. Fault lines were shown by Huzita and Sano [5].

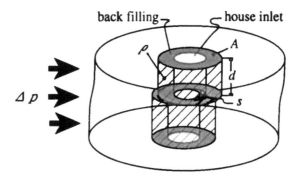

Fig. 4 Model of horizontal shear for a house inlet. Δ p : Impulsive force, A, ρ : Area and density of back filling, d : Depth between the ground surface and shear surface, s: Displacement of house inlet. The upper part moved the s at the depth, d, due to Δ p.

Figure 5 is the distribution of the impulsive force derived from displaced house inlets. The amount of impulsive force in eastern Kobe is larger than that of the western Kobe. Its larger amounts is indicated by spots and linear distributions. The spots are located on the soft sediments of old ponds, old rivers, valley floors, back mashes and margins of fluvial fan.

Some linear distributions are located also along the southwest, extension (concealed faults) of the pre-existing active faults (Gosukebashi Fault, Ashiya Fault, Koyo Fault etc.).

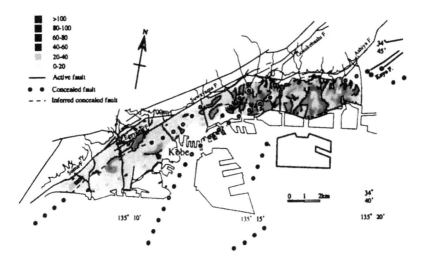

Fig. 5 Impulsive force map. The figures in the legend are the $d\sqrt{s}[cm^{1/2}]$ which are in proportion to impulsive force. Fault lines (after [5]).

SEISMIC DISPLACEMENT ON BURIED ACTIVE FAULTS BENEATH KOBE

Aki [1] discussed about seismic displacement near a fault. The component, perpendicular to the fault, of the synthetic seismogram shows a displacement pulse. This is one of the characteristic displacements due to a strike-slip fault movement. Such displacements were also recorded during the 1995 Hyogoken-Nanbu Earthquake. According to Kikuchi[7], strike-slip faulting produces the strong ground motions perpendicular to and parallel to the strike-slip fault during its approach and its arrival, respectively. The components of N40°W show pulselike shapes. Theoretically, the strike of the causative faults is N50°E and their nature is right-lateral strike-slip.

The rose diagram (Fig. 6) shows the movement direction of the house inlets. The most predominant direction of the ground move is to the northwest, and next is to the southeast, northeast and southwest. The former two directions on the diagram are perpendicular to the northeast-striking fault, while the latter two are almost parallel to the fault. Those four directions of large displacement can be explained by the theoretical interpretation by Kikuchi [7].

RUPTURE ZONE

The surface breakage appeared just on the concealed faults in the urban area. For example, one is a rupture zone appeared at southeast of Kobe University [8]. The other is a rupture zones formed along the southwest extension (concealed fault) of the Gosukebashi Fault, in the central Kobe [9]. The breakage is characterized by (1) too complicated distribution pattern, because there are hard RC structures and buildings, (2) short extension of 100-200 m long, and (3) a broad zone on the order of 200-300 m wide, and (4) a generally right-lateral displacement, on the order of a few ten centimeters to a few centimeters.

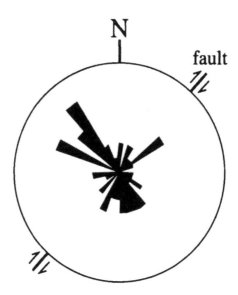

Fig. 6 Orientation of displacement of house inlets. The house inlets included in it are the impulsive force levels which are more than 100 [cm$^{1/2}$].

The same rupture zone was also recognized on the already-existing Gosukebashi Fault (active fault) in the Rokko Mountains [3, 11].

CONCLUSIONS

1. Linear distribution of the large displacement obtained from utility poles and house inlets appeared along the southwest extension (buried segment) of the pre-existing active fault.

2. Four large seismic displacements derived from the utility poles are perpendicular to and parallel to the northeast-striking fault. These are thought to have been caused by right-lateral strike-slip fault movement beneath Kobe.

3. Rupture zone also appeared along the southwest, extension (buried segment) of the pre-existing active fault.

Judging from those features, we believe that the linear distribution of surface breakage was produced by right-lateral movement on the urban active faults beneath Kobe. Numerous remarkable damage of RC structures (e.g., buildings, elevated highways and railways) is thought to have happened along the urban concealed faults of Kobe. Those surface breakage is significant in understanding the subsurface fault movement.

Acknowledgments

We wish to thank Professor Emeritus Kazuo Huzita of Osaka City University for his encouragement. This work was partly supported by the fund from Kobe University.

REFERENCES

1. K. Aki, Seismic Displacement near a Fault. *Jour. Geophys. Res.*, 73, 5359-5376.(1968)
2. H. Endo, S. Watanabe, M. Makino, Y. Murata, K. Watanabe, And A. Urabe, Subsurface geological structure with special reference to concealed faults and basement structure in Kobe and Ashiya cities, Hyogo Pref., Japan. *Butsuri-Tansa*, 48, 439-450 (in Japanese with English abstract).(1995)
3. M. Hirano, And T. Fujita, Geological Hazards from the "Hyogo-ken Nambu" Earthquake, with reference to the slip landform along the active faults. *Earth Science*, 49, 77-84 (in Japanese with English abstract).(1995)
4. K. Huzita, And Y. Maeda, *Geology of the Suma district*. Quadrangle Series, Scale 1:50000, Geol. Surv. Japan, 10lp (in Japanese with English abstract).(1984)
5. K. Huzita, and M. Sano, The Great Hanshin-Awaji Earthquake Disaster and Rokko Movements; "Earthquake Disaster Belt" associated with Buried Fault Scarps. Kagaku *Science*, 66, 793-805 (in Japanese).(1996)
6. K. Irikura, Causative Faults, Strong Ground Motions and Damages from the 1995 Hyogo-ken Nanbu Earthquake. *Butsuri-Tsnsa*, 48, 463-489 (in Japanese with English abstract).(1995)
7. M. Kikuchi, A shopping trolley seismograph. *Nature*, 377, 19. (1995)
8. T. Miyata, and Y. Maeda, Ground displacement in the urban area of Kobe during the Southern Hyogo Prefecture Earthquake. *Mem. Geol. Soc. Japan* (submitted) (in Japanese with English abstract).(1995)
9. T. Miyata, and Y. Maeda, Buried active faults inferred from co-seismic ground displacement in Kobe. In Nakagawa, K., Akamatsu, J. And Nirei, H., eds., *The Hansin-Awaji Earthquake Disaster*, 135-146 (in Japanese).(1996)
10. K. Nakagawa, Basement structures and seismic hazards. In Nakagawa, K., Akamatsu, J. And Nirei, H., eds., *The Hansin-Awaji Earthquake Disaster*, 257-269 (in Japanese).(1996)
11. T. Nakai, and H. Sunouchi, Damages along Gosukebashi Fault in the Hyogoken-Nanbu Earthquake. *The Proceedings of Symposium on The Great Hansin-Awaji Earthquake and its Geo-environments*, 69-74 (in Japanese with English abstract).(1995)
12. T. Nakata, K. Yomogida, J. Odaka, T. Sakamoto, K. Asahi, and N. Chida, Surface Fault Ruptures Associated with the 1995 Hyogoken-Nanbu Earthquake. *Jour. Geogr.*, 104, 127-142 (in Japanese with English abstract).(1995)
13. H. Nemoto, H. Kato, E. Suzuki, And K. Irikura, The aftershock distribution during the first 10 hours after the 1995 Hyogo-ken Nanbu Earthquake. *Programme and Abstracts, Seism. Soc. Japan*, 1995, No.2, A34 (in Japanese). (1995)
14. Y. Nigauri, and T. Miyata, Subsurface horizontal shear from deformed house inlets in Kobe during the 1995 Southern Hyogo Prefecture Earthquake. *Jour. Geol. Soc. Japan* (submitted) (in Japanese with English abstract).(1995)
15. S. Tanaka, and T. Okimura, Distribution of damaged houses and Topography in the eastern Kobe - Hanshin area. The Hyogoken-Nanbu Earthquake and Disaster of Topographical Condition, Kokon-shoin, 82-94 (in Japanese).(1996)

Proc. 30th Int'l. Geol. Congr., Vol. 5 pp. 143-161
Ye Hong (Ed)
© VSP 1997

Geodynamic Model for Midcontinent Resurgent Activity and Instability of India Shield

G.C.NAIK, A.K.SRIVASTAYVA AND V.N.MISHRA
KDM Institute of Petroleum Exploration ONGC, Dehradun 248195, India

Abstract

The recent earthquake activities within the stable and seismically inactive Indian shield has raised serious doubts on the rigidity of the Peninsular India. Integrated analysis of various geological/geophysical data clearly reveals the presence of two major ENE-WSW trending transcurrent faults i.e Narmada-Son-Jorhat Transform (NSJT) and Tungabhadra-Balasore-Laisong Transform (TBLT) systems which tectonically divides the Indian plate into three microplates. Each microplate appears to have distinct stress distribution pattern and is moving northeastward with different velocity.

It is envisaged that the along-strike variation of orogenic resistance and the slab pull forces, resulting from oblique collision and tectonic wedging at the NE India, play an important role in initiating the regmatic shear for the development of these continental transform faults (CTF) along preexisting shear/weak zones. The model, besides explaining the midcontinent resurgent tectonic activity, predicts that the microplate bounded by these two continental transform faults, is more susceptible to resurgent activities.

Keywords: geodynamic model, midcontinent resurgent activity, India Shield

INTRODUCTION

The Peninsular India has long been considered as a rigid and undeformed massif since the end of Precambrian time. It is therefore, regarded as one of the most stable aseismic shield of the World. Quite contrary to this widely held view, the neotectonic activities and the associated phenomenon in this shield, are really puzzling the earth scientists. Similar neotectonic activities are also widely documented in many other shield areas of the World. These activities are controlled by the preexisting geological features and therefore grouped under the general term 'resurgent tectonics' [22] . However it is not clearly known that how these tectonic phenomena are related to the plate boundary forces. In other words, a cogenetic model explaining the resurgent tectonic activities within the global tectonic framework , is yet to be developed. In India, these are being attributed to the resistance of the Indian plate to its sliding under the Asian Plate [51,52]. This is mainly based on the assumption of a rigid Indian plate moving in N to NE direction. The present paper attempts to explain the midcontinent resurgent activities and brings out their relation with overall geodynamics of the Indian plate. The study, besides proposing a different geodynamic model for the Indian plate, highlights the critical role played by the along-strike variation of the plate tectonic forces resulting from the oblique collision and tectonic wedging at Northeastern India, in initiating continental transform faults along preexisting weak zones/crustal heterogeneities.

CONTEMPORARY GEODYNAMICS AND STRESS DISTRIBUTION IN INDIA

Geologically, Indian Peninsula represents a mosaic of different tectonic/structural provinces variously named as Precambrian Mobile Belts (PMB). These are made up of high grade metamorphic rocks and are crisscrossed by a number of lineaments/faults (Fig. 1). Reactivation of these older weak/shear zones in the recent years are known by the occurrence of major earthquakes, hot springs and other associated phenomenon.

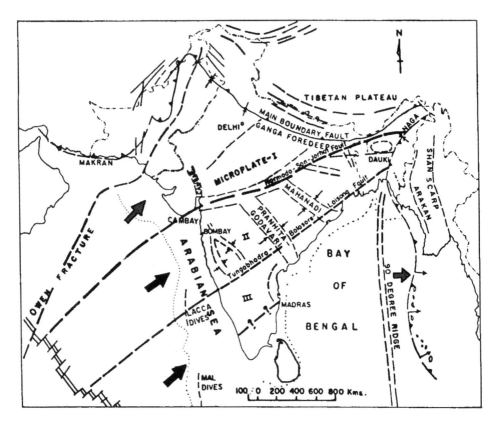

Fig. 1 Tectonic map showing the continental transform faults and microplates of India

Occurrence of earthquakes in India are more common along the Himalayan Orogenic belt which constitutes a well defined seismically active hazard zone. On the other hand, the number of earthquakes within the Peninsular India are relatively less (Fig. 2) .The recent (September 30th , 1993) earthquake (M 6.4, Depth 7km) in the Latur area, Mharastra, claiming more than 30,000 lives, was the most devastating one[39]. This is being attributed to the reactivation of the Kurduwadi rift[52]. The Koyna earthquake (M 6.4, depth 11-14km) is also the result of reactivation of the fault delimiting a 50-km wide rift valley [26]. The Pattern of seismicity (Fig. 2) displays some linear trends in NW-SE, ENE-WSW and NE-SW direction coinciding with the preexisting weak zones viz Godavari rift, Maharadi rift, Damodar rift etc. The focal mechanisms of these earthquakes show the dominance of strike slip or thrust faulting. some earthquake epicenters are also associated with a combination of strike slip and thrust faulting [8] . Based on the focal mechanism solutions, Chandra[8] suggested that Peninsular India may by under a state of left lateral shear along NNE trending vertical planes.

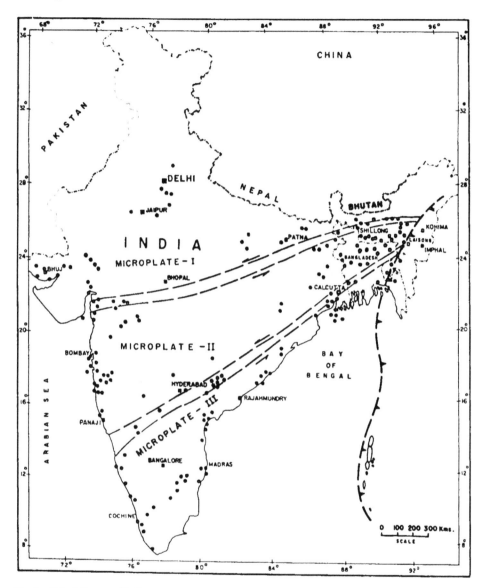

Fig. 2 Seismicity map of India

Weissel et al. [55] observed the existence of N-S, NW-SE, and E-W oriented compressive stresses throughout the Indo-Australian plate. Occurrence of thrust faults at shallow depths in Indian and Australian continental region suggest existence of large horizontal stresses [13]. The global pattern of tectonic stress analysis indicates the existence of NE trending compressive stress in the Peninsular India [60]. In situ stress measurements by hydrofracturing at different places also show maximum compressive stress in NE direction.

GEODYNAMICS OF INDIAN PLATE

Division into microplates

Separation of Greater India from the Indo-Australian-Antarctica Plate took place around M-2 time i.e. Early Cretaceous [3,15,25,38]. Subsequently, India moved northwestward [29]. This resulted into the opening of the Indian ocean in one side and the progressive closure of the Tethyan sea on the other (Fig. 3). Consequently, the northwestern corner of India collided with the Baluchistan (part of the Eurasian plate) around 50 Ma. This was the first phase of the Himalayan Orogeny (C-1 in Fig. 3) during which velocity of the Indian plate dropped from 18-19.5 cm/yr to 4-5 cm/yr [25] and the activity along the Ninety East Ridge (NER) diminished. The Mid-oceanic Ridge activity in Arabian Sea was intensified and by A-21 time (48 Ma) the NER became dormant.

The next phase of orogeny took place along the Northeastern edge of the Indian plate (C-2, Indo-Burma orogeny) during Late Eocene to Early Oligocene time[15,33]. The collision resulted into the evolution of the Arakan-Yoma Thrust complex (AYTC) and the development of the Upper Assam foreland basin [34]. During Miocene, this foreland basin experienced the terminal phase of development with the formation of Naga Schuppen Belt (NSB). In the south the Surma Basin remained as a Remnant Ocean Basin (ROB) which was later progressively closed due to the southward advancing orogenic front (Fig. 4). Bay of Bengal and subduction zone along the Sunda and Andaman Trench represents present day ROB. Thus, an Oblique Collision and Tectonic Wedging Model (Fig. 4) analogous to that of South America and Caribbean Terranes [44], has been envisaged for the Northeastern India. Figures 5. and 6. reveal the present day geodynamic scenario and the crustal sections in Northeastern India to argument this fact.

The Tethyan Sea in the North was finally closed during Late Miocene giving rise to the Himalayan Belt (third phase of Himalayan orogeny (C-3). At present the entire Himalayan belt including the Northeastern India is undergoing A-subduction (Ampferer type) which is amply evidenced by the contemporary tectonic activities in this belt.

PLATE TECTONIC FORCES

The plate movement is ultimately determined by a complex interaction of a set of forces operating at the plate boundaries [10,13,19,20,21,47,58]. Broadly these forces can be grouped as (i) Driving forces and (ii) Resistive forces (Fig. 7).

Driving Forces: The main driving forces are Slab Pull and the Ridge Push.

Slab Pull (Fsp) is exerted by the descending lithospheric slab of the Indian plate along the Andaman-Nicobar-Java-Sumatra trench and is directed normal to the strike of the trench. This force is proportional to the thickness L of the subducted slab. $Fsp \propto L3$ [30]. This force is also dependent on the convergence rate Vc [30,42] and the dip angle of the descending slab as $Fsp \propto Vc \sin$[12]. Theoretically slab pull is found to be proportional to the length of the trench and the length of the subducted slab but independent of the subducting velocity[59].

Ridge Push (Frp) results due to the gravitational sliding of Indian plate off the flank of Carlsberge ridge. It is considered as the integrated value of horizontal pressure gradient over the distance from the ridge [27]. It acts normal to the strike of the ridge and is proportional to

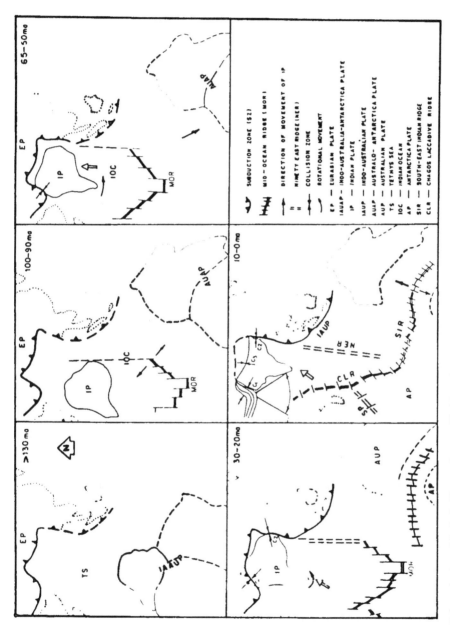

Fig. 3 Schematic view of the India Plate motion (modified from Currey et al.[15]). Note the anticlockwise rotation of IP, tectono-chronology of Himalayan orogeny (C_1, C_2, C_3) and progressive closing of India Ocean towards south

the length of the ridge. Yamano and Uyeda [59] assumed this force as independent of the
spreading rate.

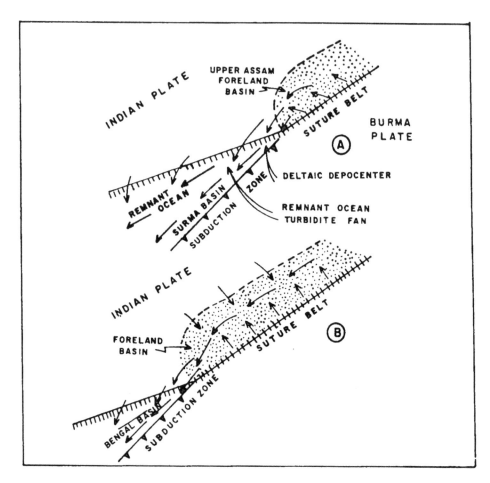

Fig. 4   Oblique Indo-Burma collision and tectonic wedging in North-Eastern India. Progressive advancement of
orogenic front (A to B) with formation of remnant ocean explains the sychronous deposition of molasse and flysch
sediments

Resistive Forces (Fr): This includes all those forces that resist the movement of the plate.

Trench resistance (Ftr)   is the resistance at the trench/subduction zones at the northern
and eastern boundary of Indian plate and results due to compositional buoyancy (Fcb)
created by the petrological stratification of the oceanic lithosphere and shearing force (Fsh)
acting along the contact between the down going Indian plate and the over riding Burmese
plate and along the boundaries of the subducted slab in the upper mantle.

Drag force (Fcd) or Shear Stress (d) acts at the base of the Indian plate and is proportional
to the area and the absolute velocity of the plate.

Suction Force(Fsu) acting on the overriding Eurasian plate (Tibetan and Burmese) segment.

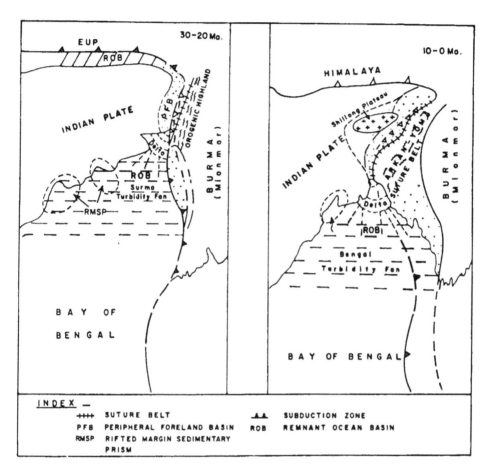

Fig. 5 Generalized paleographic map of North Eastern India and adjoining areas during Neogene time showing oblique collision and progressive closing of India Ocean

Orogenic Resistance (For) is the net resistance resulting from the collision of Indian plate along Himalayan and Indo-Burma orogenic belts.

Besides all these forces, there are also some forces acting on the lithosphere which induce local stress[50]. The ridge push and the transform fault resistance are smaller on an average by a factor of 3 to 5 [17]. Basically, the plate is driven by slab pull and ridge push and being resisted by basal asthenospheric drag and other resistances. Once collision is set in, the slab pull is lost [59] and the orogenic resistance generated due to the buoyancy constraints of the subducting slab becomes dominant.

All these forces can be combined in the form of a simplified equation. Assuming the mantle drag on the bottom of the plate as the only resistive force, the equilibrium equation can be written as:

$$Fsp + Frp + Fcd = 0 \tag{1}$$

As Fcd is proportional to the absolute velocity of the plate (V), equation (1) becomes

$$uv = Fsp + Frp \tag{2}$$

Where u is a constant coefficient for a plate of constant area.

Considering slab resistance force (Ftr), the equation (1) can be modified as

$$Fsp + Frp + Fcd + Ftr = 0 \tag{3}$$

Since Ftr is proportional to V, then Ftr = -vV.

So equation (3) becomes

$$(u + v)\,V = Fsp + Frp \tag{4}$$

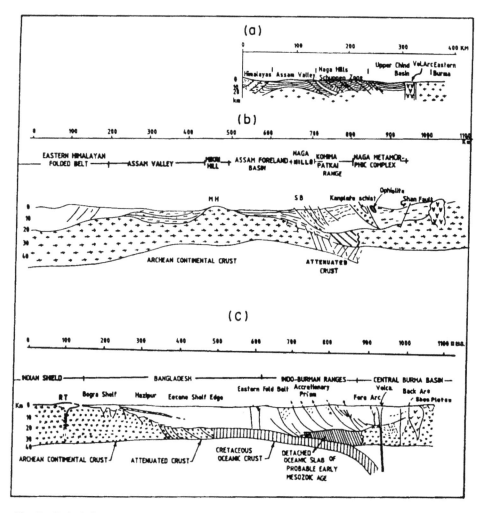

Fig. 6 Geological sections across Assam and adjoining areas, showing the along-strike variation of crustal nature and closure of sea in northeast due to oblique collision and tectonic wedging.

EFFECT OF OBLIQUE COLLISION AND TECTONIC WEDGING ON PLATE MOTION

Once the collision is set up the slab pull force (Fsp) acting on the part of the trench diminishes and the plate subducts at a slow rate. Imposing this condition to eq. (2), the velocity of the colliding block (Vcb) can be obtained by

$$u\ Vcb = Frp \tag{5}$$

and the velocity of the subducting block Vsb is given by

$$u\ Vsb = Frp + Fsp \tag{6}$$

Now dividing eq. (6) by eq. (5) we get

$$\frac{Vsb}{Vcb} = 1 + \frac{Fsp}{Frp} = 1 + a \cdots \tag{7}$$

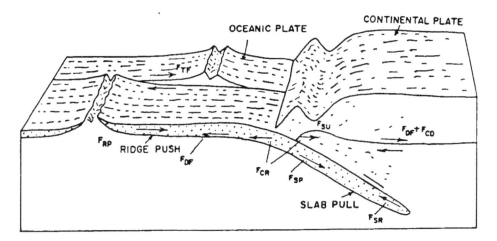

Fig. 7 Possible force acting on the lithospheric plates F_{DF}, mantle-drag, F_{CD}, extra mantle-drag beneath continents, F_{RP}, ridge-push F_{TF}, transform fault resistance, F_{SP}, slab-pull, F_{SR}, slab resistance on the descending slat as it penetrates the asthenosphere, F_{CR}, colliding resistance acting on the two plates in equal magnitude and opposite directions, F_{SU}, a suctional force that may pull the overriding plate towards the trench. (From Forsyth and Uyeda[19].)

Where 'a' represents the strength ratio of slab pull to ridge push and is generally 3 to 5 [17]. Thus Vsb is greater than Vcb by a factor of 4 to 6. In other words, after collision the initial velocity of subducting plate drops by 4 to 6 times. This is consistent with the velocity drop of Indian plate from 18-15cm/yr to 4-5cm/yr after the collision with the Eurasian plate [25]. However, the effect of oblique collision is more complicated (Fig.8). The segment where the Fsp remains operative still retains its dynamism and moves eastward relatively at a higher velocity. This creates a differential velocity field and ultimately leads to the segmentation of plate and forms a new microplate boundary. The nature of this faulted boundary depends on its relative orientation and the linear velocity vector Vr between the two blocks [23]. The preexisting crustal weak zones are more susceptible for such reactivation.

CONTINENTAL TRANSFORM FAULTS IN INDIAN SHIELD

A transform fault is a fault with a dominantly lateral (horizontal) movement that terminates abruptly at a spreading ridge, Himalayan type mountain range (located within, rather than at the edge of, a continent), or trench (subduction zone) [57]. Though these faults are present

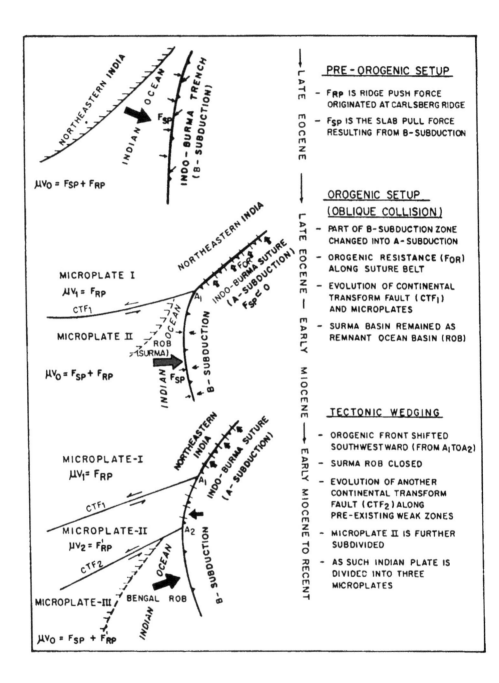

Fig. 8   Evolution of continental transforms and microplates in India Shield

areas like Circum-Pacific region, California [14], Southern Alaska[35], New Zealand [36], Eastern United States[2] and are referred as the Continental Transform Faults (CTF). Two major CTFs (Ridge-Mountain Transform Fault type) (Fig. 1) segmenting the Indian plate into three microplates, has been inferred in the present study.

Out of these two, the Narmada-Son Lineament (NSL) is well known as a major geofracture across the Indian shield. It is considered as a tectonic rift reactivated since Precambrian time[11]. Bhave et al.[4] interpreted NSL as a system of quasi parallel deep seated faults interconnected by second order faults in anastomosing pattern representing a continental transform of sinisteral wrench developed in a transtensional tectonic milieu'. Extension of MSL up to the Dauki tear fault, along the southern margin of Shillong plateau, has been suggested by many authors [5,6,18,43]. Aeromagnetic data revealed the probable extension of NSL towards Murray ridge in the west and Himalayan syntaxis in the east[31]. Biswas[6] considered Narmada-Son tectonic trend as a transform fault which starts from Carlsberg-Murray ridge of the Arabian Sea extends along Dauki wrench fault and finally merges with the Naga Schuppen belt. Sen [45]considered the extension of the SNL along the northern flank of the Brahmaputra Valley as a dextral transform fault and named it as Narmada-Son-Brahmaputra Transform (NABT) fault. However, in the present study, on the basis of the tectonic analysis and tectonostratigraphic synthesis i.e Oblique Collision and Tectonic Wedging in northeastern India, the NSL has been joined with the Jorhat Wrench System-a known wrench fault zone which separate the Upper Assam from the Dhansori Valley and the Surma Basin in the south. Thus, Narmada-Son-Jorhat Lineament forms a continental transform(NSJT) system that was reactivated during Miocene.

Another strike slip fault running almost parallel to and south of NSJT with similar geological and geophysical signatures has been inferred in the present study. It controls the course of river Tungabhadra in the west, the course of river Krishna in the east and follows the Eastern Ghat trend from north of Vijayawada. then it merges with Balasore-Kasinagar Lineament[53] and finally terminates within the Indo-Burma ranges along the Sonamura-Laisong Lineament. It is referred here as Tungabhadra-Balasore-Laisong Transform (TBLT) system. It also joins with the Carlsberge ridge in the west and acts as a continental transform fault dividing the south India further into two segments. The Indian Peninsula is thus segmented into three microplates (I,II, and III in Fig.1). The evidences in support of this observation are as follows.

Lineament Analysis: Landsat MSS data interpretation shows that the southern peninsular India is dominated by ENE-ESE,NE-SW and NNW-SSE to NW-SE trending regional megalineament systems (Fig.9). A system of parallel to subparallel ENE-WSW trending megalineament representing a major intra-continental rift zone (The Nilgiri-Eastern Ghat Zone) has been indicated by Rakshit and Rao[40]. Mahadevan[28] considered this major lineament as the manifestation of the sheared thrust boundary between the Archean craton and the fringing mobile belts along the eastern margin of the Cuddapah basin and along the southern and northern margins of the Archean Singhbhum Granite Craton. There are evidences of late rifting alkaline and carbonate magmatism in some of these thrust zones. Varadarajan and Ganju[53] have interpreted this ENE-WSW trending lineament as a fault zone parallel to the Eastern Ghat structural trends (Fig.9) which offsets the NW-SE trending older lineaments and the coastline. Therefore, these are younger in age and have shaped the present day East Coast margin of India.

Gravity-Magnetic data Analysis: Various gravity lines indicating regional highs and lows

154   G. C. Naik et al.

has been interpreted in the Indian Peninsula. A long wave length regional low has been observed in the Boguer gravity anomaly map over peninsular India which is attributed to the shallow depth of the asthenosphere[54]. Radial polarized <2°> MAGSAT scalar magnetic map shows this ENE-WSW trending lineament more clearly[32,46].

Seismological Studies Most of the earthquake epicenters in India lie close to ENE-WSW trending deep faults. Further the focal depth below Eastern Ghat Mobile Belt (EGMB) is indicative of the brittle-ductile transition within the crust. The Deep Seismic Sounding (DSS) sections have also revealed the presence of a zone of mid-crustal reflector indicating mantle disturbance(upwelling or thrusting) beneath EGMB[24].

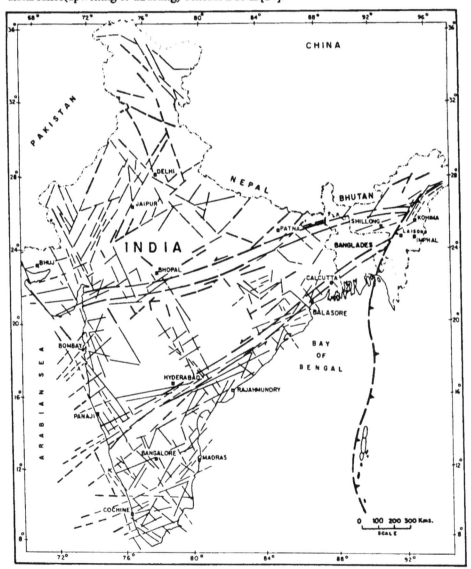

Fig.9 Lineament Map of India

Heat Flow Studies: Heat flow analysis have indicated high heat flow values along this lineament. Presence of hot sprints aligned in NE-SW direction are also known in Bihar, West Bengal and Orissa (Fig.10). Perumal et al.[37] considered this as a hot lineament(lineament associated with hot springs).

Electrical Conductivity Study have clearly brought out a high electric conductivity zone beneath the EGMB. This could be due to the relatively hotter and less viscous nature of the crust along this zone.

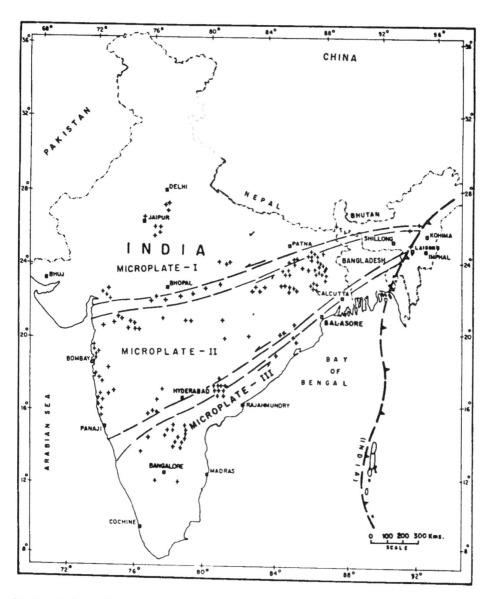

Fig. 10   Distrbution of   hot springs in India

Satellite Altimetry Data   Satellite gravity map derived from satellite altimeter data in the Arabian sea have revealed a number of lineaments [48]. Prominent trends are in NE-SW, NW-SE, ENE-WSW and NNW-SSE (Fig. 11). These lineaments represents the extension of coastal faults in deep sea. The ENE-WSW to NE-SW trending lineaments/faults are closely associated with the transform faults across Carlsberge ridge. These zones are acting as boundaries of segmented blocks and are controlling the movement of the individual blocks.

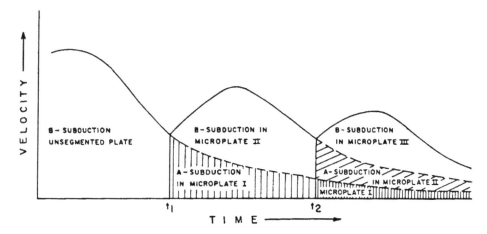

Fig.11   Schematic velocity field diagram in oblique collision and tectonic wedging setting
(Differential velocity field created due to along-strike variation of tectonic forces which in turn segment the plate into microplates.)

To sum up, various geological and geophysical evidences support the reactivation of a ENE-WSW trending weak zone as a major continental transform fault similar to the already known NSJT. Morphologically both the systems delineate broad fault zones composed of a number of faults in anastomosing pattern. The genesis/reactivation of these CTFs are attributed to the complex geodynamic processes and the operative tectonic forces at the Indian Plate boundaries.

DISCUSSION AND IMPLICATIONS

Origin of Continental Transform Faults
The oblique collision and tectonic wedging at Northeastern India during Miocene have reactivated the preexisting NE-SW to E-W trending Narmada-Son lineament into a sinistral strike slip fault which joins with the Jorhat wrench fault. Thus the Narmada-Son-Jorhat Transform (NSJT) system forms a continuous fault zone at least from Miocene. Reactivation of this fault system during Miocene has been reported by Chandra and Choudhury[7]. This fault system referred here as a major continental transform fault which separates the zone of A-subduction (the collisional belt) from the zone of B-subduction (ROB) to the south (Fig.8b). Virtually, the Indian plate gets segmented into two microplates having contrasting stress regimes caused by the fundamental plate tectonic forces operative at the boundary (microplate-I and II in Fig.8b). A triple junction of TTF(Trench-Trench-Fault) type is formed in which one of the trench boundary has been converted into a collisional boundary. The region seaward of this junction i.e the Surma Basin and its southward continuation into Bay of Bengal, possessing the trench boundary represented the remnant ocean basin(ROB) during

Miocene time[34].

With continued subduction, the orogenic front progressed further oceanward (Zipper type) and the triple junction shifted accordingly from A-1 to A-2 (Fig.8c) with time. Surma remnant ocean basin was closed during Late Miocene/Pliocene with deposition of fluvial molassic sediments and have induced development of a transcurrent fault system along Tungabhadra-Balsore-Laisong lineament within central India further segmenting the plate into another microplate. Though the movement of the orogenic front is generally slow and steady, the development of the continental transform fault is controlled by the time dependent creep phenomena and depends on the factors like crustal heterogeneity, strength and orientation of preexisting weak zones, stress generation and concentration locale within the subducting plate.

The relative velocity of the segmented microplates are schematically shown in the Fig. 12. The curves represent the ideal steady state condition. But the shape of the curve may get distorted by factors like change in boundary conditions, lithospheric heterogeneity within the plate, suction force etc. Though the velocity of the subducting plate drops after the collision, the segment where B-subduction is still going on tries to regain its original velocity. The maximum velocity attainable by the plate will be lower due to loss of some of the trench length and the ridge length with time. The subducting plate may break before attaining its original velocity. The collisional belt will slowly attain isostatic equilibrium with concomitant decrease in the rate of A-subduction and ultimately the plate becomes welded with the overriding plate. In case of Indian plate, none of the segment has reached the equilibrium condition. Even the Himalayan belt is still experiencing neotectonic activity. However, the intensity of neotectonic activity in this segment is expected to decrease with time. The segment-II is relatively more active at present. This is manifested by the devastating earthquakes of the recent times. With further southward advancement of the orogenic front i.e. with closure of the Bengal ROB, the South India will experience more seismic activity. Reactivation of another continental transform along weak zones already existing in the South India, is a likely proposition. A probable transform zone along the well known Palaghat Gap has been shown in the Fig. 1 and 12.

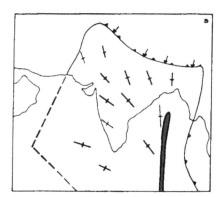 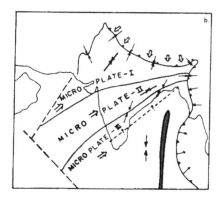

Fig. 12 Schematic stress distribution in the different microplates of India

Evolution of Micro-Plates

Along-strike variation of the plate boundary forces resulting from the Oblique Collision and

Tectonic Wedging in the Northeastern India have generated differential stress distribution pattern in and around the Indian plate. Consequently, the Indian peninsula hitherto considered as the most stable and rigid shield, is segmented into three micro-plates separated by continental transform faults (Fig.1 and 12). These major faults are parallel to and in some places directly aligned with offshore transform fracture zones of Carlsberge Ridge and therefore, define the microplate boundaries. Most of these fault zones follow the older Precambrian shear zones or preexisting weak planes. The movement along the oceanic transform faults may be getting transferred to the nearest onshore fracture zones of proper orientation. The older weakness planes controlled the placement of the transform faults in the initial stages of Indian Ocean development. further, the same weak zones also control the course of the continental transform faults within the Indian Shield. Though in the present study, the microplate-I has been considered as a single subplate, there could be further smaller divisions within this. This is because of the fact that the Himalayan orogeny as a whole is not straight forward orthogonal type. Rather it proceeded from northwest to the southeast direction. Therefore, unless the exact tectono-chronologic evolution of the Himalayan orogenic belt is worked out in detail, further subdivision of microplate-I seems difficult.

Mid-Continent Resurgent Activity and it's relation to Plate Tectonic Forces
This study brings out that the plate motion generally changes with concomitant arrangement of the plate kinematics particularly when unsubductible materials (either continental or very young oceanic crust) reach an active subduction zone. The oblique collision belts are more prone to such dynamic reorganization. Along-strike variation of the crustal nature provides the unique stress distribution pattern. The impact of this differential stress distribution can be transmitted to the deep interior of the adjacent continental crust in various forms and can trigger the midcontinent resurgent activities. Earthquakes nucleate at the intersection of the preexisting zones of weakness with the continental transforms.

Cloetingh and Wortel[13,14] have inferred an high level of regional stress field in Indian plate caused by the net boundary forces acting on the plate (Fig.12a). However, the orogenic resistance along the Indo-Burmaorgenic belt is not considered in their model. Therefore, incorporation of the resistive force must   modify the stress orientation within the Indian plate (Fig.12b) which can explain the present day tectonic activity within Indian Shield more logically. The major fracture zones as established above act as stress guides and define the direction of movement of the individual microplates. The stress distribution pattern vary from one microplate to the other. Accordingly, the stability of the different segments of the Indian plate also varies. It is reasonably speculated that the lateral compressive forces which reactivate the preexisting weak zones/crustal heterogeneities within segment-1 and II, are derived from the orogenic resistance in the Himalaya and the Indo-Burma orogenic front. The orogenic resistance from the Indo-Burma Orogenic front is more effective in segment-II. This explains the NW-SE trending chain of gravity highs as seen within the central India. Mahadevan[28] attributed these features to crustal upwarps and also to high density material deeper down. Agrawal et al.[1] favored crustal upwarping caused by the Himalayan collision tectonics. However, the NE-SW to E-W compression as envisaged in this paper, explains these features more logically. Reactivation of the NW-SE trending faults bounding the older rift systems like Mahanadi rift, Godavari rift, Maradiwada rift etc. could be to NE-SW to E-W compression rather than N-S compression from the Himalayan sector.

The subduction pull at the Andaman-Nicobar-Java-Sumatra trench and the ridge push from the Indian ocean ridge are the basic driving forces for the block-III. Neotectonic movements in this block display pattern similar to that of Indian Ocean and therefore, appear to operate under the same tectonic regime.

Reorganisation Of Indian Plate: A Futuristic View

It stems from the above discussion that the microplate-I and II are neotectonically more active Now. However, microplate-I will be attaining isostatic equilibrium with concomitant decrease in resurgent activity. Finally, it will be welded to the Eurasian plate. On the other hand, microplate-II and III will be more destabilized with further shifting of the orogenic front towards south. This leads support to the view that Arab is now a part of the Indian plate as evidenced by meager activity along Owen Fracture Zone[56]. However, only microplate-I instead of the Indian plate as a whole, seems to be getting welded to the Eurasian plate. This interpretation is conforming to the kinematics of the plate boundary forces and the relatively less activity at the Owen Fracture Zone is attributed to the orogenic resistance at the Indo-Burma ranges.

Conclusion

The Indian plate has been segmented into three prominent micro-plates bounded by two NE-SW trending continental transform faults. The dynamic interaction of the various plate tectonic forces acting at the Indian plate boundary, resulting from the oblique collision and tectonic wedging at the Northeastern India, are the main causative for reactivation/initiation of these faults along older zones of weakness/crustal heterogeneities. The relative differential movements of these microplates are strongly influenced by the subduction related forces at the Indo-Burma orogenic belt and its southward continuation to the existing Andaman-Nicobar-Java-Sumatra trench. Further, the observed resurgent tectonic activities within the Indian shield mostly along the NW-SE trending paleo-rift/weak zones are also caused by the orogenic resistance force already set in the Northeastern India.

REFERENCE

1. P. K. Agrawal, N. K. Thakur, and J. G. Negi, A deep structural ridge beneath central India. *Geophys. Res. Lett.*, 13(5), 491-494.(1986)
2. J. P.Barosh, Neotectonic movements, earthquakes and stress state in the Eastern United States. *Tectonophysics*, 132, 117-152.(1986)
3. J. Besse, And V. Courtillot, Paleogeographic map of the continents bordering the Indian Ocean since the Early Jurassic. *Jr. Of Geophy. Res.*, 93, B10, 11791-11808.(1988)
4. K.N. Bhave, J. L. Ganju, And Jokhan Ram. Origin, Nature an Geological significance of Lineaments. *Memo. Geol. Soc. Of India.* No.12, pp. 35-42.(1989)
5. S. K. Biswas, Tectonic framework and evolution of the western margin basins of India. *Indian Jr. of Pet. Geol.* 1(2), 276-292. (1987)
6. S. K. Biswas Tectonic framework and evolution of the graben basins of India. *India Jr. Of Pet. Geol.* 1 (2), 276-292.(1993)
7. P.K. Chandra, And L.R. Chowdhary, Stratigraphy of Cambay basin. Bull. ONGC, 6. pp.
8. U. Chandra, 1977. Earthquakes in Peninsular India, A Seismotectionic study. *Bull. Seis. Soc. Amer.*, 67, 1387-1413. (1969)
9. , U. Chandra, Seismicity, earthquake mechanisms and tectonics along the Himalayan mountain range and vicinity. *Phy. Earth and Planet. Int.*, 16, 109-131.(1978)
10. W.M. Chapple, And T.E. Tullis, Evaluation of the forces that drive the plates. *J.Geophys. Res.*, 82, 1967-1984.(1977)
11. V.D. Chaubey, Narmada-Son lineament, India. *Nature Phys. Sci.*, V.232., pp. 38-40.(1971)
12. S. Cloetingh, And R. Wortel, Regional stress field of the Indian plate. *Geophys. Res. Lett.*, 12, 77-80.(1985)
13. S. Cloetingh, And R. Wortel, Stress in the Indo-Australian plate. *Tectonophysics*, 132, 49-67.(1986)
14. J. C. Crowell, The San Andreas fault system through time. *Jr. Geol. Soc. Lond.*, 136, 293-302.(1979)
15. J. R. Curray, F. J. Emmel, D.G. Moore, And R.W. Raitt, Structure, Tectonics and geological history of the

northeastern Indian Ocean. *The Ocean Basins and Margins: The Indian Ocean*. 6: (ed. A.E.M. Narain and F.G.Stehli) Plenum Press, New York, 399-450.(1982)

16. B. Das, and N.P. Patel, Nature of Narmada-Son Lineament. *Jr. Geol. Soc. India*. 25(5), 267-276.(1984)

17. G.F. Davies, The role of boundary friction, basal shear stress and deep mantle convection in plate tectonics. *Geophys. Res. Lett.*, 5, 161-164.(1978)

18. P. Evans, The tectonic framework of Assam. *Jr. Gel. Soc. India*. 5, 80-96.(1964)

19. D.W. Forsyth, And S. Uyeda, On the relative importance of driving forces of plate motion. Geophys. *J.R. Astron. Soc.*, 43, 163-200.(1975)

20. R.G. Gorden, A. Cox, and C.E. Harter, Absolute motion of an individual plate estimated from its ridge and trench boundaries, *Nature*, 274, 752-755.(1978)

21. J.R. Harper, On the driving forces of plate tectonics Geophys. *J.R. Astr. Soc.*, 40, 465-474.(1975)

22. W.J. Hinze, W.L. Braile, G.R. Keller, E.G. Lidiak, Models for Midcontinent Tectonism. In *Continental Tectonics, Studies in Geophysics*, National Academy of Science, Washington D.C. p. 73-83.(1980)

23. B.E. Hobbs, W.D. Means, P.F. Williams, *An outline of Structural Geology*. John Wiley and Sons, Inc. 571p.(1976)

24. K.L. Kaila, and Hari Narain, Crustal structure along Kavil-Udipi profile in the Indian Peninsular shield from deep seismic sounding. *Jr. Geol. Soc. India*, 20(7), pp. 307. (1979)

25. C.T. Klootwijk, J.S. Gee, J.W. Peirce, G.M. Smith, and P.L. McFadden, An early India-Asia contact: Paleomagnetic constraints from Ninetyeast Ridge, ODP Leg 121. *Geol.*, 20, 395-398. (1992)

26. N. Krishna Brahman, and J. Negi, Rift valleys Beneath Deccan Traps (India). *Geoph. Res. Bulletine*, 11 (3), 207-237. (1975)

27. C.R.B. Lister, Gravitational drive on oceanic plate caused by thermal contraction. *Nature*, 257, 663-665. (1975)

28. T.M. Mahadevan, Comments on some major Tectonic lineaments in Indian, Their Geophysical expression and Metallogenic significance. *Memo. Geol. Soc. Of India*. No.12, pp. 43-48. (1989)

29. R.G. Markl, Further evidences for the Early Cretaceous break up of Gondawana, *Marine Geology*, 26, 41-45. (1978)

30. D.P. McKenzie, Speculations on the consequences and causes of plate motions. *Geophys. J.R. Astron. Soc.*, 18, 1-32. (1969)

31. D.C. Mishra, Possible extension of Narmada-Son lineament from Murrey Ridge in the west to the eastern syntaxial bend in the east. *Earth and Planet. Sci. Lett.* 36, 301-303. (1977)

32. D.C. Mishra, Satellite magnetic map and tectonic correlation. *Jr. Geol. Soc. India*, 28, 301-303. (1986)

33. D.N. Mitra, and S. Ghosh, The other sides of the problem. *Memo. Geol. Surv. Ind.*, 119, 104-106. (1986)

34. G.C. Naik, Subsurface Geology and Tectonosedimentary Evolution of Premiocene Sediments in Upper Assam. Unpublished Ph.D thesis, Indian School of Mines, Dhanbad. (1994)

35. F.P. Naugler, and J.M. Wageman, Gulf of Alaska: magnetic anomalies, fracture zones and plate interactions. *Bull. Geol. Soc. Am.*, 84, 1575-1585. (1973)

36. R.J. Norris, R.M. Carter, and I.M. Turnbull, Cainozoic sedimentation in basins adjacent to a major continental transform boundary in southern New Zealand. *Jr. Geol. Soc. Lond.* 135, 191-205. (1978)

37. N. V. A. S. Perumal, S. G. Tewari, and T. M. Mahadevan, Delineation, discrimination and mapping of lineaments from Landsat analysis and their correlation with Aeroradiometric and Geophysical data. In Regional Geophysical Lineaments Their Tectonic and Economic Significance. Ed. M.N. Qureshy and W.J. Hinze, *Geol. Soc. India*, Memoir 12, 155-164. (1989)

38. Powell, cMca., S.R. Roots, and J.J. Veevers, Pre-break up continental extension in East Gondawanaland and The early opening of the Eastern Indian Ocean, *Tectonophysics*, 155, 261-283. (1988)

39. B. P. Radhakrishna, A tragedy too deep for words: Latur earthquake of 30th September 1993. *Jour. of Geol. Soc. India.* 42, 523-526. (1993)

40. A. M. Rakshit, and P..P. Rao, Megalineaments on the face of the Indian Subcontinent and their geological significance. In Regional Geophysical Lineaments Their Tectonic and Economic Significance. Ed. M.N. Qureshy and W.J. Hinze, *Geol. Soc. India, Memoir* 12, 17-24. (1989)

41. B. K. Rastogi, Seismotectonics Inferred from earthqukes and earthquake sequences in India during the 1980 s. *Current Science*, 62, 101-108. (1992)

42. F.M. Richter, and D.P. McKenzie, Simple plate models of mantle convection. *J. Geophys.*, 44, 441-471. (1978)

43. R.K. Roy, and R.N. Kacker, Tectonic analysis of Naga Hills orogenic belt along eastern peri-Indian structure. *Him. Geol.*, 10, 376-402. (1980)

44. R.M. Russo, and R. C. Speed, Oblique collision and tectonic wedging of the South American Continent and

Caribbean terranes, *Geol.*, **20**, 447-450. (1992)

45. N. Sen, The Narmada-Son -Berhmaputra transform: a Mesozoic fracture zone in the Gondwanic India. *Tectonophysics*, **186**, 359-364. (1991)

46. B.P. Singh, MAGSAT in Lineament studies: Results from Indian region. *Memo. Geol. Soc. of India*. **12**, 181-188. (1989)

47. S.C. Solomon, N.H. Sleep, and R.M. Richardson, On the forces driving plate tectonics: Inferences from absolute plate velocities and intraplate stress. *Geophys. J.R. Astron. Soc.*, **42**, 769-801. (1975)

48. A.K. Srivastava, D.S. Mitra, R.P. Agarwal, T.J. Majumdar, K.K. Mohanty, and B. Sahai, Gravity linears from Geosat Altimetric measurments and their tectonic implications in Arabian Sea and coastal zones of India. *Abstract. 30Th annual convention and seminar on Space applications in Earth System Science, Indian Geophysical Union, Hydrabad*. p.3. (1993)

49. S. Stein, and E.A. Okal, Seismicity and tectonics of Ninetyeast Ridge area: evidence for internal deformation of the Indian plate. *J. Geophys. Res.*, **83**, 2233-2246. (1978)

50. D.L. Turcotte, and E. R. Oxburgh, Stress accumulation in the lithosphere. *Tectonophysics*, **35**, 183-199. (1976)

51. K.S. Valdiya, *Aspects of Tectonics: Focus on South-Central Asia*, Tata-McGraw Hill, *New Delhi*, pp. 319. (1984)

52. K.S. Valdiya, Latur earthquake of 30 September 1993: implications and planning for hazard-preparedness. *Current Science*, **65** (7), 515-517. (1993)

53. K. Varadarajan, and J.L. Ganju, Lineament analysis of Indian Peninsula and the Himalayas. D. Reidel Pub. Co., Dordrecht, also Allied publisher, Madras, pp. 61-91. (1989)

54. R.K. Verma, Gravity field, seismicity and tectonics of India Penisula and the Himalayas. D. Reidel Pub. Co., Dordrecht, also Allied publisher, Madras, pp61-91. (1985)

55. J.K. Weissel, R.N. Anderson, and C.A. Giller, Deformation of the Indo-Australian plate. *Nature*, **287**, 184-291. (1980)

56. D.A. Wiens, S. Stein, C. Demets, R.G. Gorden, and C. Stein, Plate tectonic models for Indian Ocean "intraplate" deformation. *Tectonophysics*, **132**, 37-48. (1986)

57. J.T. Wilson, A new class of faults and their bearing on continental drift. *Nature*, **207**, 343-347. (1965)

58. R. Wortel, and S. Cloetingh, A mechanism for fragmentation of oceanic plates. *Am. Assoc. Pet. Geol., Mem.*, **34**, 793-801. (1983)

59. M. Yamano, and S. Uyeda, Possible effects of collision on plate motion. *Tectonophysics*, **119**, 223-244. (1985)

60. M.L. Zoback, and others, Global patterns of tectonic stress: A status report on the World stress map project of the International Lithosphere Program. *Nature*, **341**, 291-298. (1989)

Proc. 30⁰ Int'l. Geol. Congr., Vol. 5 pp. 163-169
Ye Hong (Ed)
© VSP 1997

Volcanic Phenomena in the Iwo-jima Island

NORIO YOSHIDA
National Research Institute for Earth Science and Disaster Prevention, Ibaraki, Japan Masahisa Kakiuchi
Department of Chemistry, Faculty of Science, Gakushuin University, Tokyo, Japan

Abstract

Gamma-ray surveys at ten points in the Iwo-jima Island were carried out using a NaI(Tl) detector. The gamma-ray spectrum was obtained with a multichannel pulse height analyzer. Strong intensity of the gamma-rays ^{214}Bi, an offshoot of inert gas ^{222}Rn, was observed at the northern part of the island as expected, which is active area. The data of gamma-ray survey indicated the existence of fluid transportation at the northern part of the island. It was proven that the gamma-ray survey method were useful for estimating the activity of a volcano.

The stable isotope ratios of hydrogen and oxygen, tritium concentration, and concentration of chlorine ion in the volcanic water from the island were measured. The heavy δ D, δ ^{18}O, low tritium and chlorine ion concentration were obtained. The data obtained indicated that the origin of the volcanic thermal waters from the Iwo-jima Island should be ascribed to the deep seawater, and that many micro-earthquakes occurring at the Iwo-jima Island are caused by the flow, expansion, or vibration of fluid body in the vapor-liquid separation process of seawater which is being repeated at the depth of underground.

Keywords: gamma-ray survey, volcanic activity, Iwo-jima Island

INTRODUCTION

The Iwo-jima Island is located at 24°45 to 48' N and 141°17 to 19' E, approximately 1,250km south to the mouth of Tokyo Bay (Fig. 1). This volcanic island is located in the southern part of the Izu-Mariana Arc, which includes two active volcanoes of Moto-yama (MOT) and Suribachi-yama (SUR). Gases erupt and volcanic water pours out in each corner of the island. Seismic activity usually occurs three or four times a day on the average [2]. When the seismic activity is outstanding, the seismic activity jumps to about 8 times, and approximately 30 times of earthquakes occur [2].The scales of these earthquakes are at magnitudes from 0 to 0.2, and most of them fall within magnitude 1. In almost all cases, the earthquakes have hypocenters within 2.5 km of epicentral distance at a depth shallower than -2 km [1].

A gamma-ray survey on the Iwo-jima Island was performed for the purpose of detecting the relation between the activity of the fluid upwelling rate and that of the active area. The spectral gamma-ray survey is capable of picking out the characteristic gamma-rays emanating from ^{214}Bi which is an offshoot of ^{222}Rn in the uranium decay series with a half-life of 3.825 days. The uranium decay series is shown in Fig. 2. Since ^{222}Rn is one of the inert gases, a ^{222}Rn concentration measurements is advantageous in monitoring the upwelling rate of the fluid [4]. In addition to this property, because ^{222}Rn has a rather short half-life, the ^{214}Bi concentration, which can be measured with a gamma-ray detector, is equivalent to the ^{222}Rn concentration, i.e., the activity of the fluid upwelling.

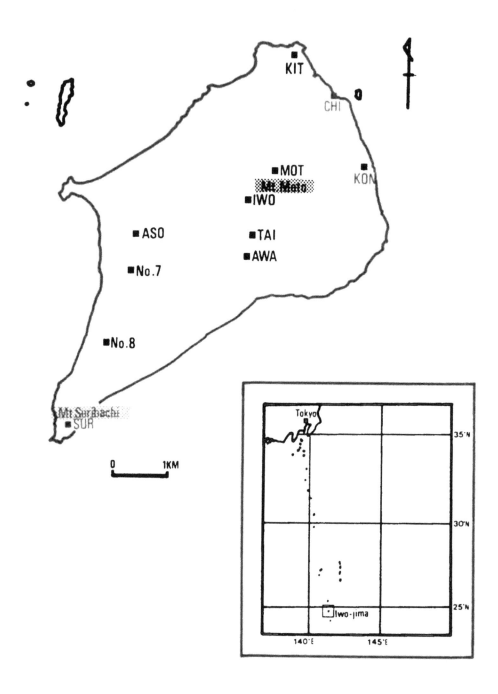

Fig. 1 Location of the Iwo-jima Island and gamma-ray survey points

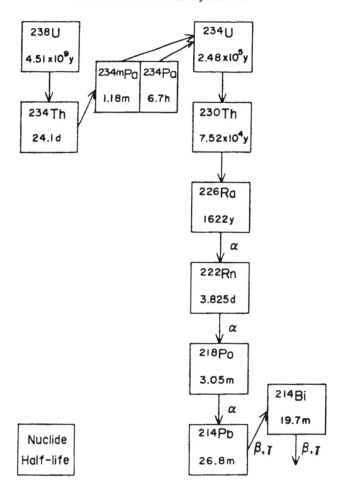

Fig.2 Radioactive decay series including ^{222}Rn

The stable isotope ratios of hydrogen and oxygen, tritium concentration, and concentration of chlorine ion in the volcanic water from three points were measured for the purpose of estimating the volcanic phenomena occurring underground of the Iwo-jima Island.

GAMMA-RAY SURVEY

The example of a gamma-ray spectra and results of gamma-ray survey are shown in Fig. 3 and Fig. 4, respectively. The figure shows the following interesting fact. The northern part of the island had a relatively large ^{222}Rn concentration in volcanic water. That is, ^{222}Rn emanations at MOT were larger than that at SUR. We understand that the volcanic gas with ^{222}Rn, the parent of ^{214}Bi, must be transferred upward MOT from the deep underground of the Iwo-jima Island.

Upper: IWO;Lower:No. 8. IWO is located at the active area. No. 8 is located at the non-active area.

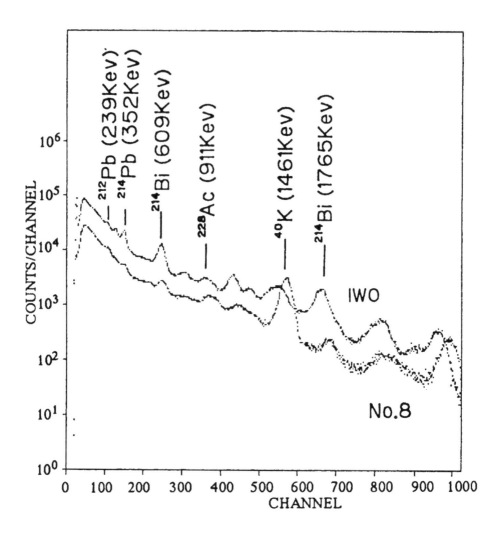

Fig. 3 Example of gamma-ray spectra observed at the Iwo-jima Island.

δD, δ ^{18}O, TRITIUM AND CHLORINE ION CONCENTRATION

The data of δD, δ ^{18}O,, tritium and chlorine ion concentration are shown in Table 1. The stable isotope ratios of hydrogen and oxygen in the volcanic water collected from four points on the island are; δ D=+32.2 to +42.5 ‰, δ ^{18}O=+8.7 to +11.3‰, respectively. As a result of comparison with rainfall onto the Iwo-jima Island, δ D=-8.0‰, δ^{18}O=-1.6‰, as well as surface seawater δ D=+0.9‰ and δ ^{18}O=0‰, the D and ^{18}O of the volcanic water showed remarkable concentrations. The tritium concentration in rainfall onto Iwo-jima Island and that in surface seawater were 5.3 T.U. and 4.3 T.U., respectively. In comparison with them, the respective tritium concentrations of the volcanic water showed low values at 0 to 2 T.U..Further, in comparison with the chlorine ion concentration of approximately 563 meq/l of surface seawater around the Iwo-jima Island, the volcanic water showed extremely low values from 0.2 to 6.8 meq/l of chlorine ion concentrations. Deep seawater, surface seawater, rainfall, and magmatic water can be considered to be the factors forming the origin of the volcanic water from the Iwo-jima Island, however, the extremely low tritium concentration in the volcanic water from Iwo-jima Island indicates that neither the rainfall nor the surface seawater can be the origin of the volcanic water. Moreover, when the light $\delta D = -40$ to -80 ‰ of the magmatic water is taken into account, it is difficult to explain the extremely heavy δD obtained at the Iwo-jima Island. In consequence, it is the most appropriate to think that the origin of the volcanic water should be ascribed to the deep seawater.

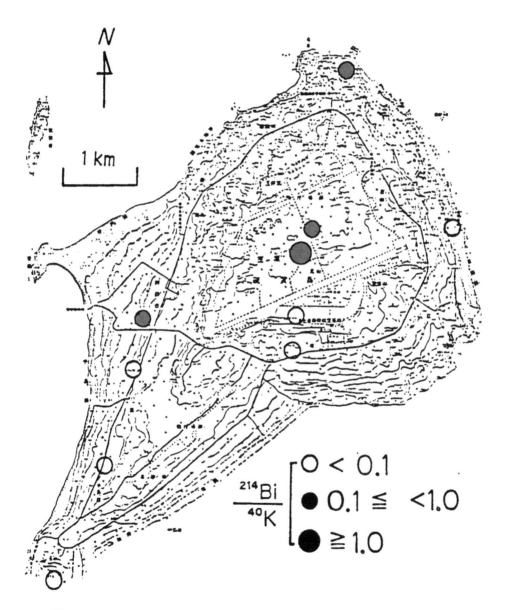

Fig.4 ^{222}Rn emanation rate at the Iwo-jima Island.

The vapor-dominated system suggested by White et al. [3] is an important concept to understand the volcanic phenomena occurring underground of the Iwo-jima Island.In order to explain the heavy δD, δ^{18}O and a low chlorine ion concentration of the volcanic water, we thought of a volcanic water model as described the next: The seawater that is taken into the deep underground of the Iwo-jima Island undergoes concentration of D and ^{18}O due to the interaction between rock and seawater. The vapor phase that undergoes vapor-liquid separation by geothermal energy concentrates both the D and the ^{18}O, and the vapor phase is condensed again. In so doing, the condensed phase extremely decreases its chlorides. It is assumed that after this process of evaporation and condensation is repeated several times under the ground, the volcanic water of the Iwo-jima Island pours out of the ground surface. We suppose that many of the micro-earthquakes occurring at the Iwo-jima Island are caused by the flow, expansion, or vibration of fluid body in the vapor-liquid separation process of seawater being repeated at the depth of underground. As mentioned already, because the majority of the micro-earthquakes that occur at the Iwo-jima Island have their epicenters shallower than the depth of -2km, it can be assumed that the vapor-liquid separation of seawater seldom occurs at the depth of more than -2km of the Iwo-jima Island.

Table 1   The stable isotope ratios of hydrogen and oxygen, tritium concetration, and concentration of chlorine ion in the volcanic water.

Sampling Point	δ D	δ ^{18}O	T.U.	Cl(meq/l)
CHI	36.9	10.4	0 ±0.5	
ditto	35.1	9.0	2.0±0.3	6.8
KIT	33.0	8.7	1.1±0.2	
ditto	42.5	11.3	1.5±0.2	0.3
IWO	32.2	9.1		
ditto			0.4±0.2	0.2
Railfall	-8.0	-1.6	5.3±0.2	
Seawater	0.9	0.02	4.3±0.2	563

CONCLUSIONS

Relatively strong gamma-rays of ^{214}Bi were observed at northern part of the island. This is the evidence for the transportation of gas with ^{222}Rn, the parent of ^{214}Bi.

A gamma-ray survey is useful to for estimating the activity of a volcano.

The origin of the volcanic thermal waters from the Iwo-jima Island should be ascribed to the deep seawater.

Many micro-earthquakes occurring at the Iwo-jima Island are caused by the flow, expansion, or vibration of fluid body in the vapor-liquid separation process of seawater being repeated at the depth of underground.

REFERENCES

1. M. Kasahara and Y. Ehara. Eight temporal seismic observations in 1968-1978 and microtremor activity on Iwo-jima, and upheaval model, *Chigaku-zasshi*, 94, 62-71 (1985).
2. T. Kumagai.Volcanism and seismicity in the Iwo-jima volcanoes, *Chigaku-zasshi*, 94, 53-61 (1985).
3. D. E. White, L. J. P. Muffler and A. H. Truesdell. Vapor-dominated hydrothermal systems compared with hot-water systems, *Econ. Geol.*, 66, 75-97 (1971).
4. N. Yoshida and H. Tsukahara. Gamma-ray spectral survey and ^{14}C measurements on the biological communities at the subduction zone Sagami Trough using the submersible "SHINKAI 2000", *J. Phys. Earth*, 39, 255-266 (1991).

Proc. 30ᵗʰ Int'l. Geol. Congr., Vol. 5 pp. 171-181
Ye Hong (Ed)
© VSP 1997

On the Application of the M8 Algorithm to New Zealand Earthquake Data

DAVID VERE-JONES[1], MA LI[2] AND MARK MATTHEWS[3]

1)Institute of Statistics and Operations Research, Victoria University of Wellington, Wellington, New Zealand
2)Centre for Analysis and Prediction, State Seismological Bureau, Beijing 100036, China
3)Department of Mathematics, MIT, Cambridge, MA 02139-4307, USA

Abstract

This paper reviews the application of the M8 algorithm to earthquakes in the New Zealand local catalogue, reported in Ma Li and Vere -Jones (1997), and compares it to earlier applications of M8 to the Californian local catalogue reported in Keilis-Borok, Kossobokov and Rinehart (1986), Keilis-Borok, Knopoff, Kossobokov and Rotvain (1990) and Matthews and Switzer (1992)

The New Zealand local catalogue contains many smaller, especially intermediate-depth, events not included in the NEIC catalogue for which M8 was originally designed. Used with local magnitudes M_L in place of the M_s or m_b given in the NEIC catalogue, and applied to the full catalogue of intermediate and deep as well as shallow events, M8 produces 4 TIPS for vents with target magnitude $M_L \geq 7$, one uccessful, one failed, and wo current. Both regions covered by current TIPS have experienced large hallow events since the TIPS were declared, one just under and the other just at $M_L = 7$.

These results are very similar to those concerning the application of M8 to NEIC data for California reported by Keilis-Borok et al/.(1990), and confirmed subsequently in Matthews and Switzer (1992). They suggest that in both cases the algorithm leads to probability gains of about one order of magnitude. In both cases, however, the results appear to rest on some surprising details. It is pointed out in Matthews and Switzer (1992) that the results for California are critically dependent on the inclusion of a group of small events at the Geysers Geothermal area, which may have been partly induced by human activities. In New Zealand, the results are dependent on the inclusion of intermediate-depth earthquakes in the catalogue, and cannot be reproduced if these events are omitted. The possibility cannot be ruled out that the observed effects are due, in part at least, to improved network capability and reporting procedures.

Keywords: M8 Algorithm, New Zealand earthquake

INTRODUCTION

This paper summarises the key results concerning the application of the M8 algorithm to New Zealand data presented in Ma Li and Vere-Jones[5], and compares some salient features with those reported for the application of M8 to Californian data in Keilis-Borok et al.[2] and Matthews and Switzer [6].

In contrast to the global studies of large earthquakes for which M8 was originally introduced, the New Zealand and Californian studies both relate to regional studies in regions of moderate seismicity. Although both form part of the Pacific boundary, the tectonic environments of the two regions are very different. The New Zealand region covers two oppositely sensed subduction zones, each with depths reaching several hundreds of kilometres, a volcanic zone, and a   transition zone. The Californian region is dominated by the transcurrent San Andreas fault and its associated fault systems.

Despite these differences, the results reported in Ma Li and Vere-Jones[5] and Keilis-Borok et al. [4] are strikingly similar, at least if taken at face value.   The algorithm is used to seek out 5-year intervals of increased probability ("TIPS") for events of magnitude 7 or greater. In both cases the region was covered by a series of overlapping circles (7 for New Zealand and 8 for California) and the algorithm was run for a time period of around 20 years, starting around 1975. In both cases just 4 TIPS were obtained, those in California covering all three events with $M_S \geq 7$ occurring in the region during the observation period. Likewise in New Zealand , the TIPS covered   both of the two events with $M_L \geq 7$ which occurred in the region during the observation period. Granted that in each case the TIPS covered only a small proportion (around 1/7) of the total time-space interval covered by the study, this is an impressive record. It represents a probability gain of around 5-10 in each case, and is likely to have occurred by chance alone- assuming random allocation of TIPS to circles and study periods - with a probability of around 1/50 for each region individually, or around 1/2000 for both regions together.

In both cases, however, a more careful probing of the factors giving rise to the TIPS   reveals some disquieting features. Matthews and Switzer[6]   point out that the results for the Californian data are sensitive to the inclusion of a group of small events the Geysers Geothermal area which are thought to have been induced by   injection of water to enhance steam production. The results for the New Zealand data depend critically on the inclusion in the catalogue of the intermediate and deep events from the subducting plates. These show striking increases, which may be partly due to improved network coverage,   towards the end of the study period, a little before the time when two current TIPS were declared.

The paper contains three further sections. In the following section, the main features of the M8 algorithm are recalled and its application to the New Zealand local catalogue is outlined. Then the results for the New Zealand catalogue are summarised and compared with those for California. The final section is a short discussion section.

PROCEDURE

The New Zealand analysis made use of the draft software prepared by Kossobokov and others for inclusion in Vol 6 of the IASPEI software series. The basic algorithms are the same as in earlier descriptions such as Keilis-Borok and Kossobokov [4], but the draft manual gives   more explicit details of aspects such as the windowing routine used to produce the catalogue of main events and recommended procedures for dividing the observation period into training and evaluation periods. We briefly recapitulate the operation of the algorithm and the procedures adapted for the New Zealand analysis, following Ma Li and Vere-Jones [5] and the description given in the draft manual.

The aim is to forecast times of increased probability (TIPS) within prescribed regions of observation. These regions are conventionally taken as circles, with radii dependent on the magnitude (M_0) of the target event. For forecasting events with $M \geq 7$ (i.e. $M_0 = 7$) the suggested radius is 280km; for events with $M \geq 8$ ($M_0 = 8$), it is 670km. For areas of moderate seismicity, such as New Zealand and California, $M_0 = 7$ is the natural choice, and the only one that will be considered in the sequel. (An account of the $M_0 = 8$ analysis for

New Zealand is included in Ma Li and Vere-Jones [5]; it produced one false TIP for a region centered to the North-East of New Zealand, and two current TIPS. No events with $M_L \geq 8$ have been recorded anywhere in New Zealand since the beginning of the study period.) For each observation region, a catalogue of main events is prepared, using a simple space-time windowing technique to remove aftershocks. Details of the procedure are given in the draft manual. Each event in the catalogue of main shocks has three attributes: an origin time t_i, a of associated aftershocks A_i; more precisely, A_i represents the number of aftershocks, over the basic catalogue threshold, in the first two weeks following the i-th main shock. The epicentral coordinates are used to determine which events enter the catalogue for a magnitude M_i, and a number given observation region, but are not used after that.

The algorithm was intended for use with the NEIC catalogue. This presented a difficulty for the New Zealand analysis, since many smaller, especially intermediate depth, events listed in the local catalogue do not appear in the NEIC catalogue. In fact, there are insufficient NEIC data to run the algorithm successfully for   New Zealand. For this reason, the New Zealand analysis was based on data from the New Zealand local catalogue, using a version of M_L in place of the m_b or M_S listed in the NEIC catalogue. The differences in coverage of the two catalogues, and between the magnitude scales, are outlined in Ma Li and Vere-Jones [5],and in more detail in Harte and Vere-Jones [1]. The windowing algorithm supplied with the M8 software was used to prepare the list of main events and their attributes for use with the main algorithm. Because the use of intermediate and deep events was not originally envisaged for the M8 algorithm, additional catalogues were made for the shallow (d $\leq$40km) and deep (d $\geq$40km) events separately.

Taking each circle in turn, the algorithm selects magnitude thresholds L_1 and L_2, separated by at least half a magnitude, and chosen to ensure minimum average rates of 10 (respectively 20) events per half-year over the initial (training) period. (If insufficient data are available to carry out a meaningful analysis an error message appears). It then proceeds to compute a set of seven time series, each computed half-yearly. Let t denote the time in years since the beginning of the study period, $N_j(t)$ the cumulative number of main events up to time t with magnitudes exceeding L_j (j=1,2), and $S_j(t)$ the cumulative sum

$$S_j(t) = \sum e^{0.46M_i} \qquad (1)$$

where the sum is extended over all main events in the region for which $0 \leq t_i < t$ and $L_j \leq M_i \leq M_0 - 1/2$. Then the seven series may be defined as follows.

1. Two series $X_1(t)$, $X_2(t)$ proportional to six-yearly mean rates:

$$X_j(t) = N_j(t) - N_j(t\text{-}6), j=1,2. \qquad (2)$$

2. Two series $Y_1(t)$, $Y_2(t)$ proportional to departures of these from the running mean up to year t-6, $t \geq 6$:

$$Y_j(t) = X_j(t) - 6\frac{N_j(t-6)}{t-6}, j = 1,2. \qquad (3)$$

3. Two series $Z_1(t)$, $Z_2(t)$ proportional to a non-linear combination of strain release and frequency:

$$Z_j(t) = \frac{S_j(t) - S_j(t-6)}{[X_j(t)]^{2/3}}, j = 1,2. \qquad (4)$$

4. One series recording extreme aftershock sequences:

$$A(t) = \max_t A_i, \qquad (5)$$

the maximum being taken over main events (t_i, M_i) with $t-1 \le t_i < t$ and
$M_0 - 2 \le M_i \le M_0 - 0.2$.

The overall study period, say $(0,T)$, which is determined largely from considerations of network coverage and completeness, is then divided into two parts, an initial training period $(0,T_1)$ and an investigation period (T_1,T). The training period is used to determine upper percentiles for each series; these percentiles are used as thresholds for anomalous values of the corresponding series in the subsequent investigation period. The 90% percentile is used for the first six series, and the 80% percentile for the final series. A TIP is declared at some time $t > T_1$ if, at both times $t - 1/2$ and t, the final series and at least five of the six remaining series all exceed their threshold values. Once declared, a TIP lasts for 5 years. It is declared successful (STIP) if an event of magnitude M_0 or greater occurs during this period, unsuccessful (FTIP) otherwise. A TIP which is still in force is referred to as a current TIP (CTIP).

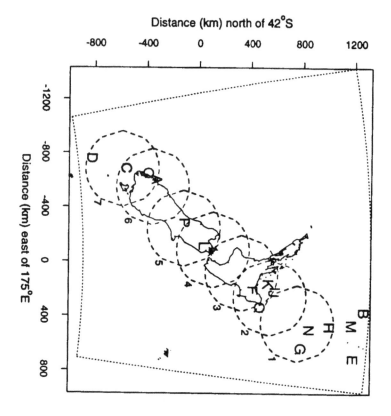

Fig. 1 Observation regions and major events from the New Zealand local catalogue. See Table 1 for a listing of events A – Q. Shaded regions show current TIPS.

The New Zealand analysis was carried out on the series of seven overlapping regions shown in Fig. 1. The total observation period was taken to be (1963-1995), of which the period (1963-1975) was used as the training period, and the remainder as the investigation period. The analysis was repeated three times, once on the full catalogue, and once each on the separate catalogues of shallow and deep events referred to earlier. Figure 1 also shows the major events to have occurred within any of the observation regions during the investigation period (1976-1995). The events are listed in Table 1.

A similar procedure, with 8 overlapping circles, was used for the Californian data (shallow events only), and we refer to Keilis-Borok et al. [3], Matthews and Switzer [6] for details.

Table 1: Major New Zealand Events ($M_L \geq 6.4$), 1976-1995

Label	Date	Location	Depth	M_L	M_s
A	4/5/1976	Fiordland	19	6.5	6.6
B	29/3/1979	Off N.E. Coast	33	6.5	-
C	12/10/1979	Off S. Coast	33	6.5	7.4
D	25/5/1981	Far S.	33	6.4	7.6
E	26/1/1983	Far N.E.	271	7.3	-
F	8/3/1984	Edgecumbe	75	6.4	-
-	12/1984-1/1985	Bay of Plenty	Swarms	5-6	5-6
G	26/9/1985	Far N.E.	33	7.0	7.0
H	13/9/1986	Far N.E.	112	6.4	-
I	25/5/1989	Bay of Plenty	259	6.6	-
J	22/3/1990	Bay of Plenty	203	6.5	-
K	16/11/1991	Bay of Plenty	264	6.4	-
L	27/5/1992	Cook Strait	84	6.7	-
M	8/8/1992	Off N.E.	395	7.1	-
N	14/12/1992	Off N.E.	255	6.7	-
O	10/8/1993	Fiordland	5	6.7	-
P	16/6/1994	Arthur's Pass	Shallow	6.5	7.1
Q	5/2/1995	Off E. Coast	Shallow	6.9	7.5

RESULTS

Table 2 summarises the results in \citeasnoun{mali} for the New Zealand local catalogue and for $M_0=7$. The original paper gives additional results for the NEIC Catalogue and for $M_0=8$.

Table 2: Summary of Results for the New Zealand Local Catalogue

Circle			All		Shallow		Deep	
No.	Lat(°S)	Long.(°E)	TIPS	Start	TIPS	Start	TIPS	Start
1	35	180	STIP	1982	FTIP	1986	FTIP	1981
							FTIP	1989
2	37	178	-	-	-	-	CTIP	1992
3	39	176	CTIP	1994	-	-	CTIP	1992
4	41	174	CTIP	1993	FTIP	1976	CTIP	1992
5	43	172	-	-	NA	NA	NA	NA
6	45	168	FTIP	1988	NA	NA	NA	NA
7	47	166	-	-	FTIP	1982	NA	NA

Notes: NA means insufficient data in the circle to run M8 successfully;
- means no TIPS in this circle during the observation period;

The STIP in Region 1 for the overall analysis covers the shallow event G ($M_L=7.0$) to the N.E. of New Zealand. The CTIP in Region 3 just covers the recent shallow event Q ($M_L=6.9$)

off the E. Coast; it has been treated as an STIP in evaluating the performance of the algorithm. The CTIP in region 4 just covers the Arthur's Pass event P (M_L=6.5) in the northern half of the S.Island. Although this was a large event, and its magnitude is still the subject of some dispute, it has not been classed as a STIP for evaluation purposes. The FTIP in region 6 covers event O (M_L= 6.7), the largest event in the Fiordland region to occur during the observation period. Although this was not counted as a STIP either, it does add to the general impression that the factors triggering the TIPS have some link with subsequent large earthquakes.

The disquieting features of the analysis relate to the shallow and deep events. Applied to the shallow catalogue by itself, the algorithm fails completely, producing no apparent link between TIPS and large events. Applied to the deep catalogue by itself, it produces three TIPS in the central region, each covering one or other of the large shallow events P and Q. All three regions show an increase in intermediate depth activity starting in the late 1980's. This certainly plays some role in triggering the TIPS. However it is not completely clear that the observed increase in deep activity is to be attributed to physical changes. A substantial upgrade in network quality occurred in 1987, when the majority of network stations were digitised. There may have been some concomitant improvement in the detection and registration of deeper events, even when such events were within the considered bounds of detectability during the earlier part of the observation period. Since the regions showing the increase are among the best covered by the network, and since the depths are not extreme, it is perhaps unlikely that the effect is purely instrumental. Figs 2-4 show a number of statistical indicators for seismicity in Regions 3,4 and 5 in the depth ranges 0-40km, 40-100km and 100-400km. The first panel in each figure shows the individual events by time and magnitude. The second shows 6-monthly counts of events over the threshold magnitude M_L=4.5. The third is a Gutenberg-Richter (log-frequency/magnitude) plot of the magnitude distribution. Systematic departure from the straight-line curve indicates problems with completeness. The fourth panel is a backward cusum (cumulative sum) plot of the magnitude. It plots values of $\sum_1^n (M_i - \overline{M})$ where $\overline{M}$ is a mean magnitude taken from data in the most recent period (1988-93). Absence of lower magnitude data, due to incomplete coverage in the early years, would show up as a change in slope from horizontal to a downwards trend from right to left. The final panel is a similar cusum for frequencies of occurrence. It may be significant that the increases are almost entirely confined to the second of these depth intervals. For this depth-range the cusum plots of the mean magnitudes tend to decrease with distance into the past, the reverse of what would be expected from any simple form of incompleteness due to lack of coverage in the early part of the period. This feature also tends to support the view that the change is real rather than an artefact of improved network capability, but the latter possibility is difficult to rule out completely.

As mentioned earlier, there are interesting parallels with the applications of M8 reported for the Californian data. The original analysis by Keilis-Borok et al. [4] used a series of 8 overlapping regions, a similar observation period, and reported 4 TIPS, each covering one of the 3 major events (Mendocino earthquake, Imperial Valley earthquake, Loma Prieta earthquake) to have occurred in California in the study period. There were no complications with deep events here, but the subsequent study by Matthews and Switzer[6]revealed a number of other disquieting features. They found that the analysis was sensitive to the choice of starting date and the exact radius used in defining the region of observation . The most

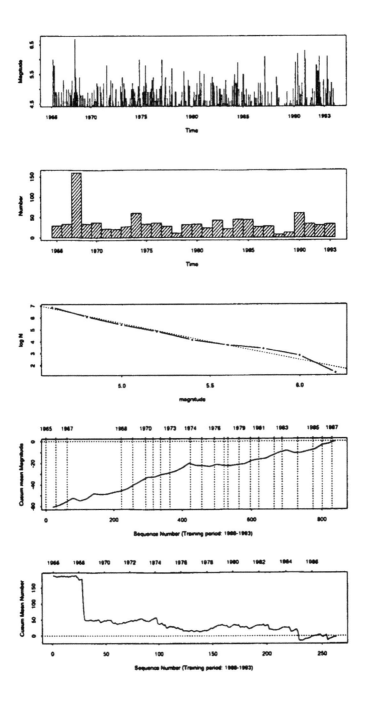

Fig. 2: Completeness study for pooled data in regions 3, 4, 5 with depths d ≤40 km. See text for further explanation.

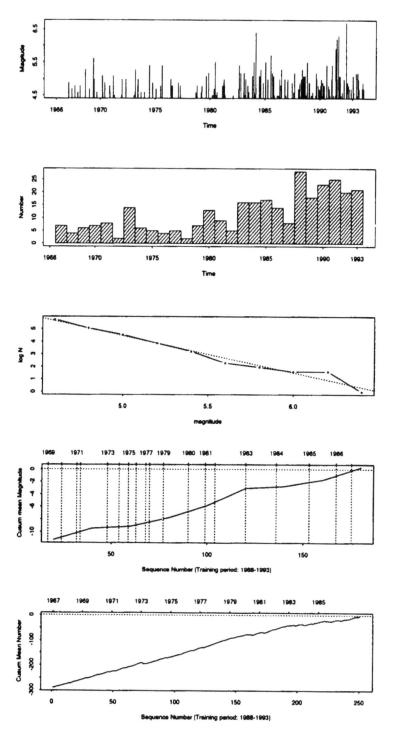

Fig. 3: Completeness study for intermediate depth events (40 < d ≤120 km) in regions 3, 4, 5. See text for further explanation.

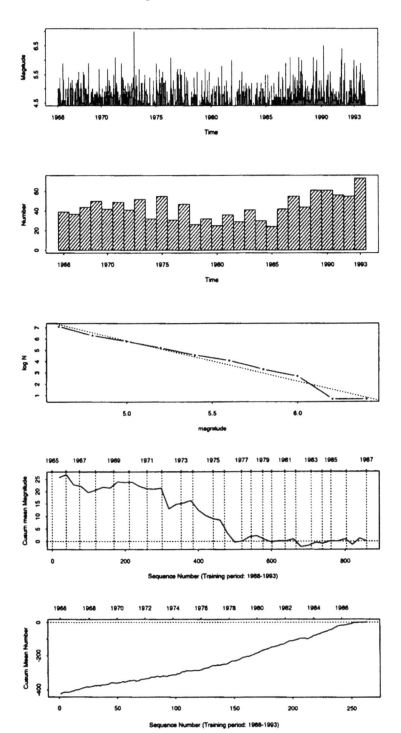

Fig. 4: Completeness study for deep events (d > 120 km) in regions 3, 4, 5. See text for further explanation.

most stable results related to the Loma Prieta earthquake, but even here small modifications to the observation regions were sufficient to alter the conclusions. For example, an important role was played by the Long Valley/Mammoth Lakes volcanic swarms near the boundary of the circle. Even more disquieting, exclusion of a small group of events from the Geysers Geothermal Region, a group thought to have been caused by injection of water for the purposes of increasing steam production, was enough to suppress the TIP. One is left with the impression that the successful TIPS for California might just be caused by a set of coincidences, without real physical meaning. That this should occur simultaneously, for two quite separate geographical regions, stretches the arm of coincidence rather far, but still not beyond the bounds of possibility.

DISCUSSION

There are still many additional studies that could be done to help clarify the issues raised by such initial analyses. In their earlier study, Matthews and Switzer looked at the sensitivity of the results to small perturbations of features such as starting date and location and radii of the circles of observation. A similar study has yet to be done for the New Zealand data. They pointed out that the seven time series used by M8 are highly correlated among themselves and suggested that standard statistical methods such as discriminant analysis could give improved ability to characterise the joint behaviour of these series associated with large subsequent earthquakes. It would seem very desirable to move away from the hard boundaries of the original algorithm, particularly in relation to quantities such as magnitude which are themselves subject to considerable uncertainties. More fundamentally, it would be desirable to incorporate the information used by the algorithm into a probabilistic model, perhaps a type of generalised linear model, which at each time period put out a probability of occurrence of a large event as a function of past and present values of the seven   series. This would permit a more comprehensive evaluation of its performance to be made, even on limited data.

The ambiguity of interpretation referred to at the end of the previous section   — whether the TIPS are triggered by real physical events or by artifacts of the earthquake registration process   — cannot be entirely removed by further analysis, however. Its ultimate source is the paucity of large earthquake data at a local or regional level, and this is likely to remain a feature of such studies for the foreseeable future. Combining studies from several regions, as hinted at in the present study, may alleviate this problem, but only partially, since each region will have peculiarities of its own that will mitigate against a joint interpretation of the data.

Acknowledgements

We are grateful to Willie Lee and the authors of Volume 6 of the IASPEI PC software series for permission to make use of their material. David Harte has helped us greatly in preparing the data sets and (with Zhuang Jiancan) the figures. This work was started under NZFRST grants VIC309 and VIC406, and is being continued as part of a collaborative research programme   supported by the ASIA 2000 Foundation of New Zealand and the State Seismological Bureau in China.

REFERENCES

1. D. Harte, and D. Vere-Jones, Differences in coverage between the NEIC and local New Zealand catalogues, Technical report, Victoria University of Wellington (1995).
2. V. Keilis-Borok, V. Kossobokov, and W. Rinehart, The test of algorithm M8 Western US, Academy of Sciences of ussr, Moscow, pp. 51-52 (1986).
3. V. Keilis-Borok, , L. Knopoff, V. Kossobokov, and I. Rotvain, Intermediate-term prediction in advance of the Loma Prieta earthquake, *Geophys, Res. Lett.* 8: 1461-1464 (1990).
4. V. Keilis-Borok, and V. Kossobokov, Premonitory activation of earthqukes flow: algorithm M8, *Physics of the Earth and Planetary Interiors* 61: 73-83 (1990).
5. Ma Li and D. Vere-Jones, Application of M8 and lin-lin algorithms to New Zealand earthquake data, *New Zealand Journal of Geology and Geophysics* 40 (1997).
6. M. Matthews, and P. Switzer, An evaluation of earthquake prediction algorithm M8 in California, Technical report, Department of Mathematics, MIT (1992).

Proc. 30ᵗʰ Int'l. Geol. Congr., Vol. 5 pp. 183-192
Ye Hong (Ed)
© VSP 1997

Load/Unload Response Ratio Theory (LURR) --A New Approach to Prediction of Earthquakes and Other Geological Disasters

YIN XIANGCHU[1], CHEN XUEZHONG[1], SONG ZHIPING[2], YIN CAN[3]

1) Center for Analysis and Prediction,SSB,Beijing 100036,China
2) Institute of Geophysics,SSB,Beijing 100081,China
3) China Disaster Reduction Press,Beijing 100029,China

Abstract.

The response to loading is different from that to unloading when the focal media is damaged since damage is an irreversible. The ratio of response rate during loading to that during unloading, called Load/Unload Response Ratio Y (LURR), could be a measure of the damage degree and closeness degree to instability and is used in a new approach to earthquake prediction. Retrospective examination of some hundred earthquake cases (from $M4$ to $M8.6$) indicates that for more than 80% of the examined ones, the value of Y is much higher than 1 for a period before the mainshock, but the Y value always fluctuates slightly about 1 during two decades for seven stable regions, so that the parameter Y could indicate a dangerous degree of an impending earthquake. More than ten earthquakes occurring in the Chinese mainland in recent years as well as the Northridge Califonia,USA earthquake (Jan. 17, 1994, $M_W6.7$) and Kanton earthquake (Japan,Sep.11,1996,$M_S6.6$) have been predicted beforehand with this method.

Keywords: earthquake prediction, LURR theory

INTRODUCTION

What is the physical essence of an earthquake? From the viewpoint of mechanics, it is exactly the damage and failure or instability of the focal media accompanied by rapid release of stress and strain energy. Let us consider the constitutive law of a fault zone or a rock mass containing pre-existing cracks (faults) or a weakened zone under high pressure and high temperature. In Figure 1(a) the ordinate denotes the general load P and the abscissa is the response R to P. At first we define two parameters as follows. One is the response rate X defined as

$$X = \lim_{\Delta P \to 0} \frac{\Delta R}{\Delta P} \qquad (1)$$

where ΔP and ΔR denote the increments of load and response respectively. Another is the Load/Unload Response Ratio (LURR) Y

$$Y = \frac{X_+}{X_-} \qquad (2)$$

where X_+ and X_- refer to the response rate under loading and unloading conditions respectively. It is well known that for the elastic phase the response rate X_+ (during loading, $DP>0$) equals to the response rate X_-(during unloading, $\Delta P<0$), hence $Y=1$ (the segment OA in Fig. 1). But $X+>X-$ ($Y>1$) happens as the load exceeds the elastic limit. The Y value becomes larger as the rock sample is approaching failure. And while it reaches point T in Fig. 1, the response rate $X+$ becomes infinite, but the value of X-does not become infinite, so that the value of Y must be infinite. The point T may be called pre-instability precursor point.

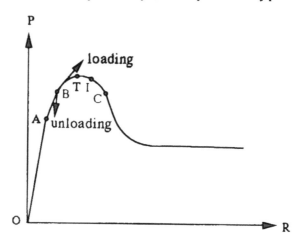

Fig.1   The constitutive law of the fault zone

From the viewpoint of damage mechanics, the seismogenic process is the damage process of focal medium which hopefully can be measured quantitatively by a damage parameter D. There are many ways in defining D. A direct way is to define D as the varying rate of modulus of elasticity M (a four order tensor) or for simplicity as the varying rate of one component of the   modulus tensor. For example, Lemaitre defined [7]

$$D = \frac{E_0 - E}{E_0} = 1 - \frac{E}{E_0} \tag{3}$$

$$D = 1 - \frac{1}{Y} \tag{4}$$

where E_0 is the Young's modulus for undamaged state of the material and E for the damaged one. Assuming the modulus for unloading condition is equal to E_0, it is easy to derive that from expressions (2) and (3). It implies from expression (4) that there is a   close relation between D and Y, so the damage degree of the focal medium can also be measured with parameter Y.

Even if D is defined otherwise, there will still be a close relation between D and Y. This suggests that the parameter Y is available for measuring quantitatively the seismogenic process and for earthquake prediction [15,17,19,22].
There are several main problems to be solved in order to predict earthquakes by means of the parameter Y. One of them is how to load and unload a block of crust and how to distinguish loading from unloading. Another one is to choose a suitable parameter as the response for LURR theory.

(1) How to load and unload a block of crust?
The linear dimension of a seismogenic zone may reach hundreds even thousands of kilometers. One of the means to load and unload it is by the earth tide. Tidal force varies periodically, so the induced stresses in the crust continue loading and unloading it

periodically.

(2) With what criterion to distinguish loading from unloading?
To distinguish loading from unloading for rock materials in the three dimensional stress state, we resort to the Coulomb failure criterion [6]. The formula in this case has been reiterated in our papers [17,15,19,22].

(3) The remaining problem is which parameters are chosen as response R for evaluating the response ratio Y.

A lot of geophysical parameters, such as the crustal deformation, water level of wells and other seismic parameters can be chosen as response for measuring Y. We are carrying out sizeable project cooperating with some excellent geophysists in China and other countries in the above mentioned fields. In this paper, mainly the Y values for seismic energy as R are presented.
The parameter Y is defined as :

$$Y = \frac{(\sum_{i=1}^{N_+} E_i^m)}{(\sum_{i=1}^{N_-} E_i^m)} \qquad (5)$$

Where E is the seismic energy, $m=0$ or $1/3$ or $1/2$ or $2/3$ or 1, the sign "+" denotes loading and "-" unloading. When $m=1$, E^m is exactly the energy itself: $m=1/2$, E^m denotes the so-called Benioff strain; $m=1/3$, $2/3$, E^m represents the linear scale and the area scale of the focal zone respectively; $m=0$, Y is equal to N_+/N_-. In this paper all the results refer to $m=1/2$. For a given region (say $2°\times2°$) and a given time window (say several months) the Y value can be determined according to expression (5). Then the dangerous degree of an imminent earthquake in this region could be estimated in terms of the Y value.

RETROSPECTIVE EXAMINATION

Using the seismic data within and outside China, we have examined comprehensively the theory of LURR with hundreds cases of earthquakes with magnitude $M4\geq4$ [3,4, 9,18]. The results of examination are satisfying and interesting. The following is the result of examination for all large earthquakes ($8>M_s\geq7$) occurring in the Chinese Mainland from 1970 to 1992.

There are 13 large earthquakes that occurred in the Chinese mainland during this period. Among them 3 events (Qinghai Earthquake, Tonghai Earthquake and Yijitaicuo Earthquake) are omitted in our analysis because of the scant data remaining 10 events available. The variation of Y value with time before all these events has been calculated. Figure 2 shows the calculated results. It can be seen from Fig.2 that for nine events out of ten, the value of Y increases clearly before the mainshock. The duration during which $Y>1$ ranges from 1 to 3 years. Theoretically speaking, the value of Y should not be smaller than one in the regime before peak stress. But if a kind of sporadic parameter to be adopted as R such as the energies of earthquakes in expression (6),it is possible for Y value to be smaller than one (e.g. Fig.2).

In addition, examinations for tens earthquakes occurring in Japan, USA and other Countries during the last few years have been carried out [10, 22].

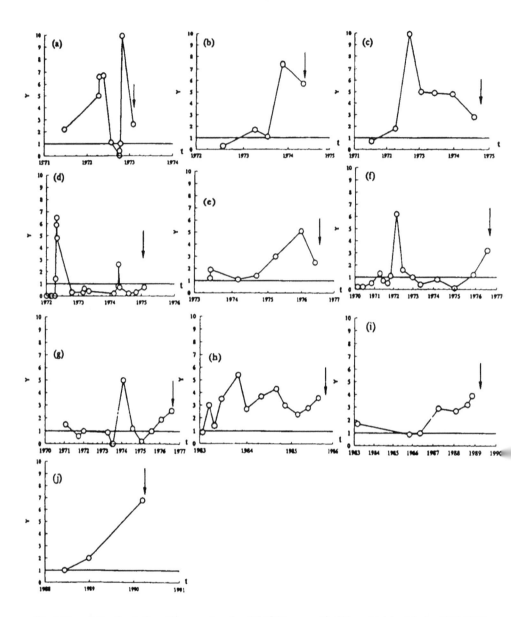

Fig. 2 Y vs relative time before all large earthquakes ($Ms \geq 7.0$) occurred in Chinese Mainland during 1970-1992.
(a) 1973,12,06 Luhuo earthquake (Sichuan Province) Ms=7.6; (b) 1974,05,11 Yongshan earthquake (Yunnan Province) Ms=7.1; (c) 1974,08,11 Wucha earthquake (Xinjiang Autonomous Region) Ms=7.3; (d) 1975,02,04 Haicheng earthquake (Liaoning Province) Ms=7.3; (e) 1976,05,29 Longling earthquake (Yunnan Province) Ms=7.4; (f) 1976,07,28 Tangshan earthquake (Hebei Province) Ms=7.8; (g) 1976,08,16 Songpan earthquake (Sichuan Province) Ms=7.2; (h) 1985,08,23 Wucha earthquake (Xinjiang Autonomous Region) Ms=7.1; (i) 1988,11,06 Lancang earthquake (Yunnan Province) Ms=7.6; (j) 1990,04,26 Gonghe earthquake (Qinghai Province) Ms=7.0

As a contrast, we chose six "stable" regions in which strong earthquakes ever occurred historically but with low seismicity in recent decades in the China mainland and analyzed their variations in Y for more than two decades (from 1970 to 1992). The results are shown in Fig. 3. Fig. 3 contrasts sharply with Fig.2. For all the six regions, their *Y* value always fluctuates slightly about 1 during the whole twenty-three years. The above results indicate clearly that the striking contrast between the variation of Y in the "stable" regions and that one before occurrence of strong earthquakes.

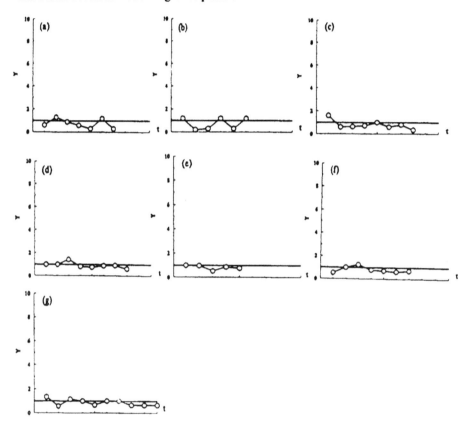

Fig. 3   Load-unload response ratio vs. time for six regions with stable crust during 1970-1992.
(a) Southern Tanlu Fault (35.5°N ± 1°, 118°E± 1°);   (b) Northern Shanxi Province (40.5°N ± 1°, 109°E± 1°);
(c) Eastern Sichuan Province (31°N ± 1°, 105°E± 1°);   (d) Northern Shandong Province (37°N ± 1°, 119°E± 1°);   (e) Western Shandong Province (37°N ± 1°, 117°E± 1°) ;   (f) Northern Henan Province (35°N ± 1°, 113°E± 1°).

It is suggested from the above results that the *Y* value indicates indeed the closeness degree to instability of a region and the LURR theory could be a new approach to earthquake prediction.

PRACTICE OF EARTHQUAKE PREDICTION

In recent years several earthquakes (M_L>6) have been predicted beforehand by means of LURR, such as the Datong earthquake (M_L=6.1, Mar. 26, 1991, in North China); the Puer earthquake (Ms=6.3, Jan. 27, 1993, Yunnan Province, Southwestern China); the Qilian earthquake (Ms=6.0, Oct. 26, 1993, Qinghai Province, Northwestern China) and the

Gonghe earthquake (Ms=6.0, Jan. 3, 1994, Qinghai Province) etc. as well as the Northridge earthquake (Mw=6.7, Jan. 17, 1994, California, USA). In Fig. 4 the Y value is plotted against time prior to the above mentioned events.

In the summer of 1993, we obtained a data set of earthquake catalogue of California for NEIC (National Earthquake Information Center, also the World Data Center A for Seismology) U.S. Geological Survey. After calculating the Y value for every region along San Anders Fault with our prediction algorithm using the data from NEIC, it has been found that the Y value increased substantially and was in excess of 1 during prolonged period (longer than one year) for 3 regions. Subsequently we mailed a letter on Oct. 28, 1993 to the scientist who offered us the dataset. It was predicted in the letter that moderate earthquakes ($7<M\leq6$) would occur probably in the three regions in one year from then on. The Northridge earthquake occurred near one of them, and the California earthquake (Sep.12, 1994, M_L=6.0) near the other two regions.

In the spring of 1996,the temporal variation of Y (LURR) for Kanto,Wakayama and Hygo regions in Japan has been calculated and analyzed. The Y value is larger remarkably than 1 near 2 years for Kanto. According to these results we predicted that an earthquake with magnitude about 6 could occurred in one year or a little longer for Kanto region (Yin,X.C. et al.,1996 which was contributed to ERC in May,1996 and published on Sep.1,1996). As expected an earthquake with Ms 6.6 occurred in this region (35.5°E,140.9°N) on Sep.11,1996.

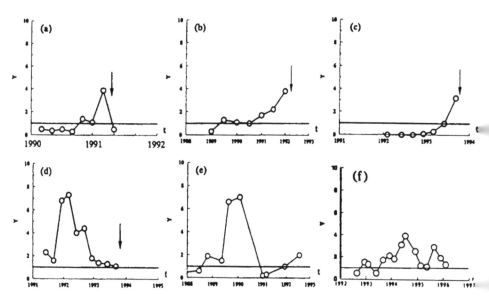

Fig. 4 Y-t charts before some cases of earthquakes predicted beforehand
(a) 1990,03,26 Datong Earthquake (Shanxi Province) M_s6.1; (b) 1993,01,27 Puer Earthquake (Yunnan Province) M_s6.3; (c) 1993,10,26 Qilian Earthquake (Qinghai Province) M_s6.0; (d) 1994,01,03 Gonghe Earthquake (Qinghai Province) M_s6.0; (e) 1994,01,17 Northridge Earthquake (California,USA) Mw6.7; (f) 1996,09,11 Chiba Earthquake (Kanto,Japan) M_s6.6

THE SPATIO-TEMPORAL EVOLUTION CHARACTERISTICS OF LURR

The spatial pattern of Y (LURR) is very complicated. Many high-Y regions exist and they form approximately a dough-nut region before the occurence of the impending strong earthquake. Most of the future strong earthquakes will be located in the dough-nut region. Take the Puer earthquake ($M \geq 7.0$) in 1979 in Yunnan Province as an example, surrounding the epicenter there exist many high-Y regions which form a dough-nut region with an area of about 25 square degrees (Fig. 5). Its stereogram looks like many peaks of mountain standing together. As a striking contrast to it, for the regions with low seismicity, the Y value of every part in it is near 1 so that the curve surface of the spatial distribution of Y will be like the topography of flat country. These characteristics could take an important role in determining the generating region of future earthquakes.

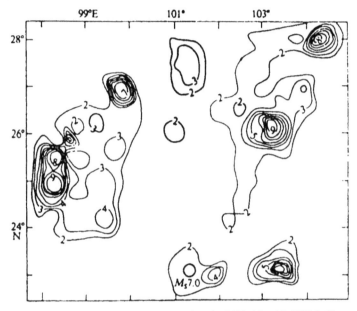

Fig. 5 The contour of *Y*2.0 one year before the Puer earthquake *M*7.0, Mar. 14, 1979 in Yunnan Province.(O stands for the epicenter)

The pattern of Y does not keep fixed and does change with time. Most of the high Y regions migrate towards to the future epicenter. It is called Convergent characteristics of high-Y regions before strong earthquakes.

We have studied 12 earthquake cases with magnitude $8 > M \geq 7.0$ in the Chinese mainland after 1970. Out of 12 studied cases, the epicenters of 11 events lie in the area where high-Y regions have converged. Only the distance between the center of the earthquake of M 7.0 of Apr. 23,1992 at the border between Myanmar and China (the boundary of its converging region) is over 50km. It is probably due to the data in the studied region is not complete since its epicenter was located in the boundary of China and Myanmar. So the epicenter of the future earthquakes could be predicted according to the above results.

In addition we have studied systematically the migrating speed of high-Y regions before strong earthquakes. It has been found that the migrating speed of high-Y regions is in order of 100 km/a, but a little different for events occurred in different regions.

Scholz (1977) has found that the speed of deformation front is about 150km per year and then Press F. and Allen C. (1995) observed that the deformation wave speed in Southern California is about 100 km per year. In this paper, the migrating speed of high-Y regions is usually the level of 100km per year. It is clear that these migrating speeds coincide with each other in order.

The data concerning Chinese earthquakes in this paper were offered by the Center of Analysis and Prediction, State Seismological Bureau of P.R.China, which is the authoritative institution of China in this field. The catalog of California came from the NEIC,USGS and the catalog of Spitak earthquake from the Institute of Physics of the Earth, Academy of Science of Russia and the data for Kanto from JMA. Since Y is the ratio of all responses in the positive period to that in the negative period, the effect of systematic error due to the magnitude of earthquakes is rather small. The minimum magnitude in the catalog used to calculate Y is as small as possible, and differs from region to region. It is M_L 2 for California, Spitak region and the Eastern China, it is equal to or even greater than M_L 3 in other regions of China.

PERSPECTIVE

As above mentioned that a lot of geophysical parameters can be chosen as response to measure Y [3,4,13]. Figure 6 shows the variation adopting the Q^{-1} of the coda wave in the Northridge region (Southern California,USA) before the Northridge earthquake (Jan.17,1994) as the response to calculate the LURR Y_Q. It can be seen that Fig.6 and Fig.4(e) coincide with each other qualitatively.

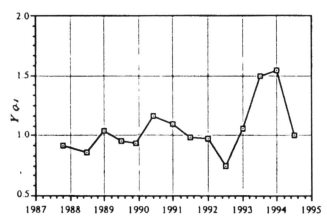

Fig.6 The variation in Y_Q prior to Northridge earthquake

To sum up, the LURR theory could open up a new approach to earthquake prediction, and it is now broadly tested in the Chinese seismological community [5]. Recently it is discovered that prior to most of the felt earthquakes (6>*M*>4) in Beijing area and even the "mine-quakes" (*M*≥2) that occurred in the Fangshan Coal Mine (in the suburb of Beijing), the Y value also increased remarkably [9]. This fact suggests that the LURR theory could be applied not only to natural earthquake prediction, but might also be utilized for the prognostication of other geological disasters such as reservoir-induced earthquake [4], rockburst, landslide, volcano eruption, etc..

Acknowledgements

The authors are very grateful to Professors Keiti Aki, Fu Chengyi, Qin Xinling, Wang Ren, Chen Zhangli, Mei Shirong, Luo Zhuoli, Zhang Guomin, Zhang Buomin, Jin Anshu, DR. E.A. Bergman, K. Hosono and H.P. Ouyang for their earnest assistance and support. This subject is supported by the Natural Science Foundation of China, the Chinese Joint Seismological Science Foundation and LNM (Lab of Nonlinear Mechanics of Continous Media) of Institute of Mechanics,SCA.

REFERENCES

1. B.K. Atkinson, The Theory of Subcritical Crack Growth with Application to Minerals and Rocks, in *Fracture Mechanics of Rock* (Ed. B.K. Atkinson) Academic Press, London (1987).
2. CHEN J.M., ZHANG Z.D.,YANG L.Z.,SHI R.H. and ZHANG J.H., Study on Variation of Aquifer Rock Parameters by Water Level Response to Earth Tide, *Earthquake*,No.1,73-79 (In Chinese with English abstract) (1994).
3. CHEN X.Z.,YIN X.C., Application of Load-Unload Response Ratio Theory to Earthquake Prediction for Moderate Earthquakes, *Earthquake Research in China*, 10 (4), 11-16 (1994).
4. CHEN X.Z.,YIN.X.C., The Changing Characteristic of the Load/Unload Response Ratio Before Reservoir-Induced Earthquakes, *Earthquake Research in China*, 10 (4) (1994).
5. DING J.H. et al., A Report on the Seismic Tendency of China of 1993, *In Investigation on the Seismic Tendency of China of 1993* (Ed. Center for Analysis and Prediction, State Seismological Bureau) (Seismological Press, Beijing 1992) pp1-68 (in Chinese) (1992).
6. J.C. JAEGER, N.G.W. COOK, *Fundamentals of Rock Mechanics*, Chapman and Hall, London, pp78-99 (1976).
7. J. LEMAITRE, How to Use Damage Mechanics, *Nuclear Eng. & Design*, 80, 233-245 (1984).
8. LI X.H., Process in Key Task Studies on Methods and Theories for Earthquake Prediction During the Time of the Eighth Five -Year Plan (Part 1) Northridge Earthquake Predicted Successfully With the Theory of Load/ Unload Response Ratio, *Recent Development in World Seismology*, (4), 24 -25, (in Chinese). (1994).
9. LIU G.P., MA Li. and YIN X.C., The Variations of Load/Unload Response Ratio Prior to Medium Earthquakes in Beijing Area, *Earthquake*, (4), 1-8 (in Chinese with English abstract) (1994).
10. Song Z.P., Yin X.C. and Chen X.Z., The time-spatial evolution characteristics of the Load/Unload Response Ratio (LURR) and its implication for predicting three elements of earthquakes, *ACTA SEISMOLOGICA SINICA*, 18, 179-186 (1996).
11. P. VARGA,.Influence of External Forces on the Triggering of Earthquakes, *Earthq. Predict. Res.*,1,191-210 (1985).
12. WANG T.W., Application of the Theory of Load/Unload Response Ratio to Earthquake Prediction with Geomagnetic Parameter. *Seismological and Geomagnetic Observation and Research* ,16,No.2,16-20 (1985).
13. YANG L.Z.,HE S.H. and XI Q.W., Study on the Variation in the Property of Rock Elasticity by Load/Unload Response Ratio of Tidal Volume Strain, *Earthquake Research in China*,Vol.10, 14.Supplement,90-94 (1994).
14. YIN X.C. and ZHENG T.Y., A Rheological Model for the Process of Preparation of An Earthquake, *Scientia Sinica (Series B)*,26,285-296 (1983).
15. Earthquake Prediction, Presented at IUGG XX, Vienna, August, also *Science in China*,38, 977-986 (1991).
16. YIN X.C., LI J., YIN C. and CHEN X.Z., Investigation of the Future Seismic Tendency of Chinese Mainland in terms of Load/Unload Response Ratio Theory, *In Investigation on the Seismic Tendency of China in 1993.* (ed. Center for Analysis and Prediction, SSB) Seismological Press, Beijing (in Chinese) (1992).
17. YIN X.C., A New Approach to Earthquake Prediction. *Preroda* (Russia's "Nature"), (1) 21-27 (In Russian) (1993).
18. YIN X.C., Review of the Progress of Load-Unload Response Ratio Theory, *Recent Development in World Seismology*, (2), 1-8 (in Chinese with English abstract) (1994).
19. YIN X.C., YIN C. and CHEN X.Z., The Precursor of Instability for Nonlinear Systems—The Load/Unload Response Ratio Theory, *In Nonlinear Dynamics and Predictability of Geophysical Phenomena Geophysical Monograph* 83 (ed. W.I. Newman and A.M. Gabrelov), Washington, pp55-61 (1994).
20. YIN X.C. et al., The Load-Unload Response Ratio Theory-A New Approach to Earthquake Prediction, *Acta Geophy. Sinica*, 37, 695-702 (1994).
21. YIN X.C.,CHEN X.Z.,SONG Z.P. and YIN C., The Load/Unload Response Ratio (LURR) Theory and Its Application to Earthquake Prediction, *Journal of Earthquake Prediction Research* 3,325-333 (1994).
22. Yin, X.C., Chen X.Z., Song Z.P. and Yin C., A New Approach to Earthquake Prediction--The Load/Unload Response Ratio (LURR) Theory, *PAGEOPH*, 145,701-715 (1995).
23. Yin X.C., Chen X.Z., Song Z.P. and Wang Y.C., The Temporal Variation of LURR in Kanto and Other Regions in Japan and Its Application to Earthquake Prediction, *Earthquake Research in China*,12,331-335 (1996).

24. ZHENG T.Y. and YIN X.C., The Subcritical Extension of Faulting and the Process of Preparation of Earthquakes, *Science Bulletin*,29,1081-1085 (1984)(in Chinese).

Proc. 30th Int'l. Geol. Congr., Vol. 5 pp. 193-201
Ye Hong (Ed)
© VSP 1997

CN Algorithm in Italy: Intermediate-term Earthquake Prediction and Seismotectonic Model Validation.

GIOVANNI COSTA[1,2], ANTONELLA PERESAN[1], IVANKA OROZOVA[1,2], GIULIANO FRANCESCO PANZA[1,2] AND IRINA M. ROTWAIN[3]

1) *Dipartimento di Scienze della Terra, Universita degli Studi di Trieste, via E. Weiss 1, 34127 Trieste, Italy*
2) *International Center for Theoretical Physics - SAND Group - ICTP, 34100 Trieste Miramar, Italy*
3) *International Institute of Earthquake Prediction Theory and Mathematical Geophysics, Russian Academy of Sciences, Warshavskoye, 79, K.2, 113556, Moscow, U.S.S.R.*

Abstractt

The CN algorithm is here utilized both for the intermediate-term earthquake prediction and to validate the seismotectonic model of the Italian territory. Using the results of the previous analysis, made through the CN algorithm and taking into account the seismotectonic model, three main areas, one for Northern Italy, one for Central Italy and one for Southern Italy, are defined. The separation among them is not marked by sharp boundaries, and on the basis of different zonations, it is possible to identify intersection areas, which can be assigned to either bordering main areas. The earthquakes occurred in these areas contribute to the precursor phenomena identified by the CN algorithm in each main area when the TIPs duration decreases when the intersection areas are included.

In a further step we have constructed a revised catalogue using the most   recent information about the seismicity in Italy, and we have considered a regionalization that follows strictly the boundaries of the areas defined in the seismotectonic model of Italy. Each of these new regions contains only the zones with similar seismotectonic characteristics. The results obtained in this way are good and stable and represent an improvement with respect to the previous investigations.

Keywords: CN Algorithm, intermediate-term earthquake prediction, seismotectonic model, Italy

INTRODUCTION

The analysis of the Time of Increased Probability (TIP) of a strong earthquake with magnitude greater than, or equal to a given threshold M_0, based on the algorithm CN, makes use of normalized functions, which describe the seismicity pattern of the analyzed area. Therefore the original algorithm, developed for the California-Nevada region, can be directly used, without any adjustment, in areas with different size and level of seismicity. The algorithm CN is described in full detail by Keilis-Borok et al.[5,6].

It has been shown [3, 4] that a regionalization, supported by seismological and tectonic arguments, leads to the reduction of the alarm duration (TIP) and of the failures to predict, and increases the stability of the algorithm compared with the results obtained when the borders of the studied area are defined simply according to the completeness of the used catalogue [7]. Therefore, the CN algorithm permits to deal with the development of modern regional geodynamic models, involving relationships between the key structural features which control the seismicity, and the selection of the optimal causative fault system for prediction purposes [12].

Considering the information contained in the seismotectonic model of Italy [10] and the
spatial distribution of the epicentres, the country can be divided into three main areas (Fig. 1)
[4]. Each of them is characterized by a dominant seismotectonic behavior, with varying
seismicity level, therefore the appropriate M_0 is used in each area.

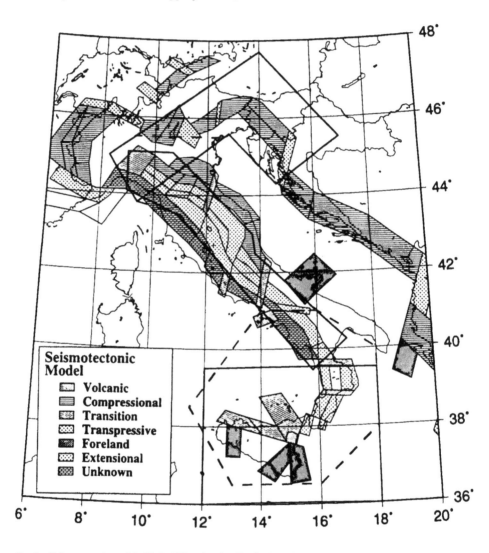

Fig. 1 Seiemotectonic model of Italy [10] and regionalization into three main and two transition areas.

We introduce here a more detailed regionalization for Nord, South and Central Italy which
follows strictly the boundaries of the seismotectonic zones [10]. Only the seismotectonic
zones with the same characteristic, and with transitional behavior between them, are
contained in each new area.

In the present analysis a new catalogue "CCI96" is routinely used for the application of the CN algorithm in Italy. The catalogue has been compiled revising the PFGING [2,3, 11] catalogue with the recently published data about the seismicity, mainly historical [1]. Some relevant differences, also for large magnitudes, have been found between the PFGING catalogue and the new CCI96 catalogue, mainly for Southern Italy where maximum magnitude is used.

REGIONALIZATION

To minimize the spatial uncertainty, the area where a strong earthquake has to be predicted, should be as small as possible, but there are three rules that limit its minimum dimensions: 1) the border of the area must be drawn following as much as possible the minima in the seismic activity; 2) the annual number of earthquakes with magnitude greater or equal to the completeness threshold of the catalogue has to be greater or equal to 3; 3) the linear dimension of the region must be about 5L to 10L, where L is the length of the expected source.

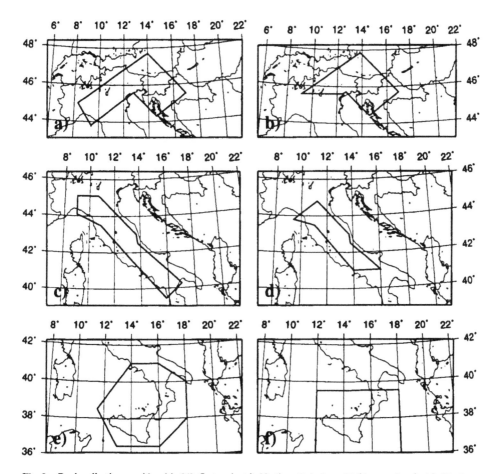

Fig. 2   Regionalization considered in [4]: first variant in Northern Italy (area 1); b) second variant in Northern Italy (area 2); c) first variant in Central Italy (area 1); d) second variant in Central Italy (area 2); e) first variant in Southern Italy (area 1); f) second variant in Southern Italy (area 2).

In the regionalization proposed by Costa et al. [4], the borders between the three main areas: Northern, Central and Southern are not sharply defined and they can be better represented by a transition domain (Fig. 1). In fact, the division of the Italian territory in three main areas, separated by two transition areas, seems to be consistent with the indications given about the properties of seismicity by the CN algorithm. In each main region, in order to analyze the effect on the prediction of the transition domains seismicity, two different regions have been tested, which blandly follow the border of the seismotectonic zones (Fig. 1, 2). In all the cases considered, the best results are obtained for the regions which include the transition areas [4].

The earthquakes information contained in ALPOR[1] has been used to construct the new catalogue, CCI96, used in the CN analysis performed in the framework of a new regionalization (Fig. 3), which follows strictly the borders of the seismotectonic zones, and in the forward monitoring in Italy.

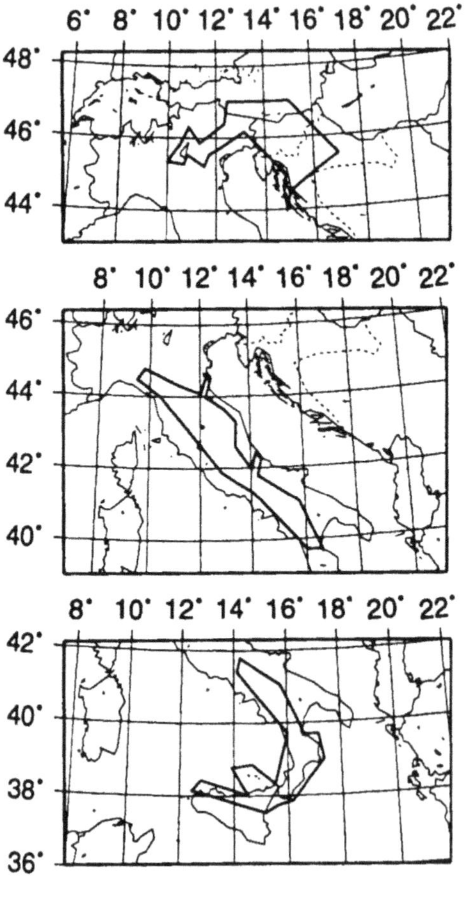

CN ANALYSIS IN NORTHERN ITALY

The Alpine arc, the most important tectonic feature in Northern Italy, is crossed by different political borders and consequently the catalogue PFGING is fairly incomplete for our purposes [4]; to fill in the gap the information contained in two other catalogues, ALPOR [1] and NEIC [9] , has been included.

According to the standards used in the CN algorithm [3], the magnitude threshold for the definition of strong earthquakes is chosen to be M_0=5.4. The period 1960-1992 is analyzed, because of the significant incompleteness of the catalogue before 1960 [4]. In the region (Fig. 2a), only 2 strong earthquakes occurred during the last 30 years (M=6.5, May 6, 1976 and M=5.4, January 2, 1988), in fact the M=6.0 September 15, 1976 event is a strong aftershock, identified as Related Strong Earthquake [8], and therefore it is not a target of the CN algorithm.

The seismogenic region, thus defined, is shown in Fig. 2a. The two strong events are predicted and the TIP duration is 27% of the total time (see Fig 4a). There is only one false alarm after the strong earthquake of 1988.

Fig. 3 Regionalization considered in the present study (solid line) Italy; b) Central Italy; c) Southern Italy; dashed lines indicate some of the variants used in [4].

In order to test the hypothesis that the earthquakes, concentrated on the edges of a tectonic structure or in the areas of intersection with other structures, cannot be neglected for the purposes of intermediate term earthquake prediction, a second regionalization (Fig. 2b), which includes only the compressive domains in the Eastern Alps (Fig. 2b), is considered by Costa et al. [4]. The two strong events are predicted (Fig 4b), but the TIP duration increases to 34% of the total time, and there are three false alarms.

The new regionalization considered here follows the compressional zones of the seismotectonic model for Northeast Italy and therefore is disconnected from Central Italy (Fig. 3a). In the Austrian and Slovenian territory the seismotectonic zones have been defined only near the border with Italy and a complete zonation is not available, therefore the border outside Italy is defined by seismicity only. The results obtained with such regionalization and using the CCI96 catalogue (Fig. 5a) are: the two strong events are predicted and the TIP duration is 28.8% of the total time with two false alarms. The reduction of the spatial uncertainty is about 28%.

CN ANALYSIS IN CENTRAL ITALY

The CN algorithm has been initially applied to Central Italy [3, 6], because the catalogue PFGING is rather complete here. Subsequently a regionalization based on seismotectonic consideration has been proposed [3] (see Fig. 2c). Only the crustal earthquakes occurred in Central Italy are used, even if, according to the model proposed by Costa et al.[4] few intermediate and deep earthquakes belong to Central Italy and should be considered when using the CN algorithm. In fact, their inclusion in the data set does not affect the results, and this is not surprising since the number of these events and their size is small. The magnitude threshold for the definition of the strong earthquakes is chosen to be $M_0=5.6$. The two strong events are predicted and the alarm occupies about 30% of the total time with two false alarms (see Fig. 4c).

The definition of the areas in Northern and Southern Italy [4] make it necessary a revision of the regionalization proposed by Costa et al.[3] for Central Italy. This revised regionalization is presented in Fig. 2d. Due to the smaller dimension of the area, the magnitude threshold is $M_0=5.4$. Four strong earthquakes occuned in the area. As it can be seen from Fig. 4d, three of them are predicted by the CN algorithm, while the 1979 event is a failure to predict; there are 4 false alarms and the TIPs increase, with respect to the previous study, from 30% to 38% of the total time.

The new regionalization (Fig. 3b) , which follows strictly the border of the seismotectonic zones, includes the extensional zones and some transitional ones. The magnitude threshold for the definition of the strong earthquakes is $M_0=5.6$ and the catalogue used is the CCI96. All the three strong events are predicted and the alarm occupies about 21% of the total time, with three false alarms (Fig. 5b).

CN ANALYSIS IN SOUTHERN ITALY

The catalogue PFGING can be considered complete in this part of Italy only after 1950, and for magnitude above 3. The magnitude threshold in the definition of the strong earthquakes, used by Costa et al.[4], is $M_0=6.5$.

198   G.Costa et al.

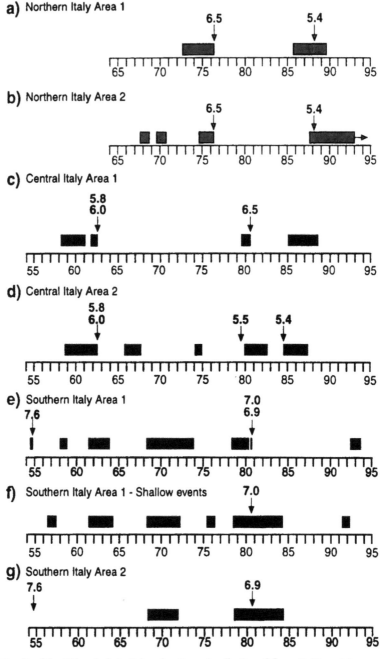

Fig. 4 Results of the CN analysis in Italy using the regionalization of figure 2. The catalogue used is the PFGING. The arrows indicate earthquakes with M≥M$_0$, TIPs are marked by black rectangles. In e) and g) the magnitude 6.9 marks an intermediate-depth earthquake in the Tyrrhenian sea; in e) and f) the magnitude 7 marks the Irpinia, 1980, earthquake and in a) and b) the magnitude 6.5 marks the Friuli 1976 event. In the Southern Italy the maximum magnitude present in the catalogue is used for the analysis, while in Northern and Central Italy the priority magnitude M$_{pr}$ (M$_L$, M$_b$, M$_d$, M$_D$) [3] is used.

a) Northern Italy

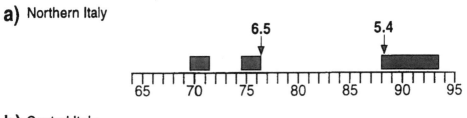

b) Central Italy

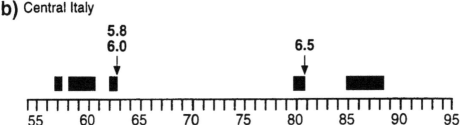

c) Southern Italy

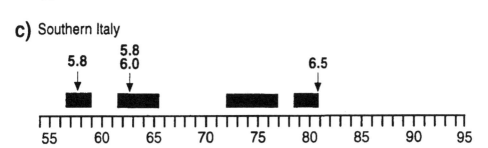

Fig. 5 Results of the CN analysis in Italy using the regionalization of figure 3. The catalogue used is the CCI96. The arrows indicate earthquakes with M≥M₀, TIPs are marked by black rectangles. The priority magnitude M_p (M_L, M_b, M_s, M_p)[3] is used.

Following the idea that the 41°N parallel divides the Apennines into two completely different tectonic domains [10], for Southern Italy the area shown on Fig. 2e has been considered. The results of the CN algorithm applied to this area [4] are reported in Fig. 4e. All three strong earthquakes (M=7.6, November 23, 1954, M=7.0 November 23, and M=6.9, November 24, both in 1980) are predicted and the duration of TIP is 33% of the total time. There are 5 false alarms.

To study the influence of the relevant deep seismicity [4] only the shallow earthquakes is considered and thus the strong event to be predicted is the M=7.0, November 23, 1980 earthquake. The diagnosis of the CN algorithm is given in Fig. 4f. The strong event is predicted, but the duration of TIP increases up to 44% of the total time and there are six false alarms.

As a second test, according to the regionalization for Central Italy [3], the northern border of Southern Italy is traced along the 39.5° parallel (Fig. 2f). In this area the two strong earthquakes to be predicted are the M=7.6, November 23, 1954 and the M=6.9, November 23, 1980 events. The 1980 earthquake is predicted with a TIP duration lasting for 25% of the total time; the 1954, M=7.6, event is a failure to predict and there are two false alarms (Fig. 4g).

The new regionalization (Fig. 3c) follow strictly the extensional and transitional seismotectonic zones present in South Italy (below the 42°N parallel). The small, but with a intense seismicity, volcanic zone present in Sicily is included as well, while the foreland seismotectonic zones have been excluded, as it was done in Central Italy in [3].

In Southern Italy the differences in the magnitude between the PFGING catalogue and the CCI96 catalogue are very large also for the events with $M>M_0$. Therefore a direct comparison with the results obtained by Costa et al. [4] is not possible. The improvement introduced by the new catalogue and the new regionalization permits to use for Southern Italy the same criteria used in Northern and Central Italy. The magnitude threshold is $M_0=5.4$ and the strong events to be predicted are 4. All of them are predicted and the duration of TIP is 31.8% of the total time (Fig. 5c). There are 3 false alarms. The spatial uncertainty reduction with respect to the results of Costa et al.[4] is relevant, about 72%.

CONCLUSIONS

The CN algorithm has been here utilized both for the intermediate term earthquake prediction and to validate the seismotectonic model of the Italian territory.

The catalogue PFGING used by Costa et al. [3, 4] has been revised using the earthquakes information contained in ALPOR [1] and a new catalogue, the CCI96, has been compiled and used here. Some relevant differences, for large magnitudes, have been found between the PFGING catalogue and the new catalogue, mainly for Southern Italy, when maximum magnitude is considered.

A new detailed regionalization for Northern, Southern and Central Italy, which follows strictly the boundaries of the seismotectonic zones [10], has been proposed. Only the seismotectonic areas with the same characteristic, or with transitional behavior, are contained in each new area. This regionalization represent an improvement of the regionalization proposed by Costa et al. [4], the average spatial uncertainty reduction of the prediction is about 45% with a general reduction of the TIPs duration and of the false alarms. The improvement of the results is particularly relevant in Southern Italy.

On the basis of the results obtained, Costa et al. conclude that the separation among the three regions proposed is not marked by sharp boundaries, and on the basis of different zonations, it is possible to identify intersection areas, which can be assigned to either bordering main areas[4]. When these intersection areas are included in the CN analysis, an improvement of the results is obtained. This result hasbeen confirmed, for Southern and Central Italy, by the results obtained using the newregionalization proposed here. In Northern Italy, the compressional seismotectonic zones are disconnected from the transitional and compressional seismotectonic zones included in the regionalization proposed by Costa et al. [4]. Therefore, a comparison is possible only with the region shown in Fig. 2b: there is an improvement with respect to the previous results, but the best result remains the one obtained when the intersection areas are included.

Acknowledgments

The authors are very grateful to Prof. V.I.Keilis-Borok for stimulating discussions. We acknowledge financial support from MURST funds, CNR-Gruppo Nazionale per la Difesa dai Terremoti contracts n.° 95.00608.54 and 96.02968.54 and INTAS grant 94-0232.

REFERENCES

1. ALPOR., *Catalogug of the Eastern Alps*, Osservatorio Geofisico Sperimentale, Trieste, Italy (computer file) (1987).
2. E. Boschi G. Ferrari P. Gasperini, E. Guidoboni, G. Smriglio and G. Valensise. *Catalogo dei forti terremoti in Italia dal 461 a.C. al 1980*. Istituto Nazionale di Geofisica SGA storia geofisica ambiente (1995)..
3. G. Costa, G.F. Panza, and I.M. Rotwain, Stability of premonitory seismicity pattern and intermediate-term earthquake prediction in Central Italy. *PAGEOPH*,144, in press.
4. G. Costa I. Orozova-Stanishkova G.F. Panza, and I.M. Rotwain. Seismotectonic models and CN algorithm: the case of Italy. *PAGEOPH*, 147 No. 1, 119-130 (1996).
5. A.M. Gabrielov, O.E. Dmitrieva, V.I. Keilis-Borok, V.G. Kosobokov, I.V. Kuznetsov, T.A. Levshina, K.M. Mirzoev, G.M., Molchan, S.K. Negmatullaev, V.F. Pisarenko, A.G. Prozorov, W. Rinehart, I.M. Rotwain, P.N. Shebalin, M.G. Shnirman, and S.Yu. Schreider, Algorithms of Long-Term Earthquakes' Prediction. International School for Research Oriented to Earthquake Prediction-Algorithms, Software and Data Handling, Lima, Peru (1986).
6. V.I.Keilis-Borok, and I. Rotwain, Diagnosis of Time of Increased Probability of strong earthquakes in different regions of the world: algorithm CN. *Phys. Earth Planet. Inter.*, 61, 57-72 (1990).
7. V.I. Keilis-Borok, I.V. Kuznetsov, G.F. Panza, I.M. Rotwain, and G. Costa, On Intermediate-Term Earthquake Prediction in Central Italy. *PAGEOPH*, 134, 79-92 (1990).
8. I., Marson, G.F. Panza, and P. Suhadolc, Crust and upper mantle models along the active Tyrrhenian Rim, *Terra Nova*, 7, in press.
9. NEIC. *World-wide earthquake catalogue*. National Earthquake Information Center (NEIC), USGS, Denver, USA. (computer file) (1992).
10. E. Patacca, R. Sartori and P. Scandone, Tyrrenian basin and Apenninic arcs: kinematic relation since late Tortonian times. *Mem. Soc. Geol. It.*, 45, 425-451(1990).
11. PFG. *Catalogo dei terremoti italiani dall'anno 1000 al 1980* (ed. Postpischl, D.), CNR-P.F. Geodinamica. Roma, Italy (computer file) (1985).
12. D.V Rundkvist, and I.M. Rotwain, Present-day geodynamics and seismicity of Asia Minor. *Computation of Seismology*, 27, (in press) .
13. I. A. Vorobieva, and G.F. Panza, Prediction of Occurrence of Related Strong Earthquakes in Italy. *PAGEOPH*, 141, 25-41(1993).

Proc. 30ᵗʰ Int'l. Geol. Congr., *Vol.* 5 pp. 203-210
Ye Hong (Ed)
© VSP 1997

Spatial Distribution of Anomalies in Different Stages before an Earthquake

MA JIN, MA SHENGLI AND LIU LIQIANG

Institute of Geology, State Seismological Bureau, Beijing, 100029, China

Abstract

A deformation cycle in a region may be divided into 4 stages, corresponding to long-term, medium-term, short-term to imminent stages before earthquake and instability, respectively. Anomalies appearing in stages I and II may be caused by changes in driving force and anomalies in stage III may be caused by fault propagating or fault weakening. In order to reveal distribution of anomalies in different stages, increments of mean stress and maximum shear stress for models with en-echelon faults and other fault patterns are studied numerically. It is shown that distribution of increment of maximum shear stress in stages I and II are similar to the background stress field while variation of incremental mean stress in stage II is outstanding at en-echelon jogs. Distribution of incremental maximum shear stress and incremental mean stress display an eight-petaline pattern and a four-quadrant pattern, respectively, surrounding the propagation area in stage III. The eight-petaline pattern, four-quadrant pattern and strong undulating of dynamic characteristics in mean stress may help to distinguish possible upcoming unstable fault.

Keywords: earthquake precursor, anomaly recognition, short-impending anomaly, numerical modelling

INTRODUCTION

A lot of observatory stations have been setting up in China since Xingtai earthquake (1966) in order to grasp earthquake precursors. Anomalies that persist for periods of a few even ten years prior to an earthquake in a region are called "long to medium-term precursors". Those appearing in periods of a few months to a few days before an earthquake are called "short-term or imminent precursors". On the other hand, anomalies appearing in vicinity of the epicenter are referred to as "source precursors", and those appearing in a rather large area as "area precursors". The complication of tectonics in China gives us an advantage to observe some anomalies near or far from the epicenter of an upcoming earthquake, which may be an alert to the occurrence of a strong earthquake. However, earthquake cases indicate that (1) there were no anomalies before some strong earthquakes, (2) in some cases no earthquake followed the anomalies, (3) places with "long to medium-term precursors" often did not correspond to places with "short-term to imminent precursors", and (4) places with obvious anomalies did not always correspond with epicenter of upcoming strong earthquake. These phenomena show that both "universality" and "regionality" in earthquake precursors may exist at the same time, which have been discussed by many authors[3-4,9-15]. We suggest that anomalies in different time scales may be caused by different mechanisms, which are related with deformation stages. Therefore, studying spatial distribution of anomalies in different stages and corresponding mechanisms may help us to understand above mentioned phenomena. In this paper, we will try to give some discussions on this topic based on mechanical behavior of rocks and some numerical results.

ANOMALIES IN DIFFERENT DEFORMATION STAGES

It is intrinsically a problem of mechanics whether anomalies appear or not. Anomalies appearing at an observatory   are attributed to the mechanical interference from changes in external mechanical condition. The interference may cause an earthquake as a result of mechanical instability in some place but only anomalies in other places. Changes in external mechanical condition may be caused by different mechanisms, such as change in external driving force, deformation of some tectonic elements easy to be deformed and propagating or weakening of faults in a region. Such an element may become   seismogenic fault in future or not. Evidence indicates that regions with heterogeneity in tectonics and spots having strong interaction with deformed elements are favored by anomalies.

In case of homogeneous fault, there is no great difference in deformation along fault,   hence no obvious precursor. However, for heterogeneous fault system   strong deformation may appear in some segments earlier, which may act as precursors of a strong earthquake. The linking of fault segments is a preparing stage for instability, and instability occurs after uniformization of sliding resistance. Therefore, anomalies appearing in both linking of faults and uniformization of sliding resistance may be regarded as "precursors".

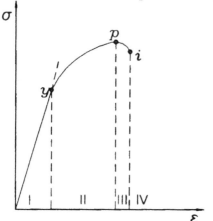

Fig 1. The characteristic points on a stress-strain curve of rock and the deformation stages

Deformation stages and corresponding anomalies can be discussed based on a common stress-strain curve of rock (Fig.1). There are three characteristic points in the curve: yield point (y), strength point (p) and unstable point (l), which may divide the curve   into four stages by them. The first one before yield point (y) is linear deformation stage   (I). The second one between point y and strength point (p) is nonlinear deformation stage (II), in which the stress field may be redistributed due to locally inhomogeneous   deformation. The third one between point p and unstable point (i) is critical stage (III), in which propagating, concentrating and linking of microcracks are predominated. The fourth one after point i is unstable stage (IV). In other words, the stages before point p (stages I and II) may be taken as long to medium-term ones. Anomalies appearing in these stages may be caused by changes in driving force or redistribution of stress field. Even though anomalies appear in these stages, instability may not occur if driven force decreases again. The stage III may be regarded as short-term to imminent ones. Anomalies appearing in this stage are related with propagating, weakening and linking of cracks. The failure may be stable or unstable depending on the ratio of stiffness in failure area to that in surrounding area. However, the failure is unreversible once the deformation enters this stage. Anomalies in this stage are closely related to upcoming stable or unstable failure.

In order to understand distribution of anomalies, stress field is divided into essential and incremental ones. The former means background stress field in a region by the action of regional driven force and the latter means variations in stress in some period caused by mechanisms mentioned above. Consequently, incremental stress field is a disturbed field superimposed on essential one. In case of absence of anomalies, the regional stress field may be regarded as being in steady state. Superimposition of disturbed stress field will break the state and cause anomalies. Therefore, incremental field can indicate clearly distribution and mechanism of anomalies.

GEOMETRIC TEXTURE MODELS AND MECHANICAL MODELS

A natural fault system is usually inhomogeneous both in geometry and in material, which may affect evolution of stress field, hence anomalies. In this paper, we focus on effect of geometric texture neglecting that of material. We chose en-echelon faults, the most common fault geometry, as geometric models and study distribution of disturbed field in different stages caused by different mechanisms including change in driving force, redistribution of regional stress field, fault propagating and so on.

Considering that a natural fault zone may be nonlinear in geometry, material and boundary[1,5-7], the triple-nonlinear method is used to deal with deformation along fault. The mechanical behavior of faults is considered as a contact problem and is simulated by a contact-impact algorithm, which permits gaps and sliding along material interface. Fault gouge is considered as a elasto-plastic material with five mechanical parameters including elastic module E, Poisson ratio v, yield limit σ, hardening module Et and hardening ratio β. Surrounding rock is considered as an elastic material with two parameters including elastic module E and Poisson ratio v. In calculation, fault propagating is achieved by changing some elements from elastic to elasto-plastic behavior, and fault weakening by decreasing the modules of fault gouge.

In this paper, define extensional stress as positive and compressional stress as negative, and maximum shear stress $\tau_{max} = (\sigma_1 - \sigma_2)/2$, and mean stress $\sigma_m = (\sigma_1 + \sigma_2)/2$. Essential fields of maximum shear stress and mean stress are expressed by the model name with τ and σ_m (e.g. EC5τ, SDQ2σ_m), respectively. Correspondingly, disturbed stress fields are expressed by $\delta\tau$ and $\delta\sigma_m$.

DISTURBED STRESS FIELD CAUSED BY DIFFERENT MECHANISMS

Essential Stress Field

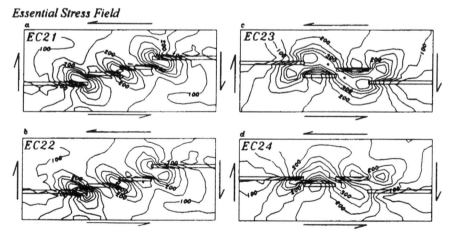

Fig. 2 The maximum shear stress field for models with different complex en-echelon faults

Four models with complex en-echelon faults are designed to study this problem. EC21 and EC22 are models with the same type of en-echelon faults and EC23 and EC24 with different type of en-echelon faults. All models are applied by the left- lateral shear in direction parallel to the fault strike and right-lateral shear in direction perpendicular to the strike. The maximum shear stress fields for these models are shown in Fig.2. For convenience, distances between two inner ends of en-echelon faults in direction perpendicular and parallel to faults are indicated by d and s respectively[8]. Take s as positive for overlapping en-echelon jog and negative for separating jog. It is clear from the figure that stress distribution in similar textural elements is similar, and stress level is dependent on d and s of jog and its position in model. For example, all jogs (1, 2 and 3, see Fig.4) are left steps in model EC21, where jog 2 with smallest d-value is of highest level of maximum shear stress. In model EC22 (Fig.2b), d-values in jog 3 and 2 are the same, but stress level in jog 3, close to border of the model, is higher than that in jog 2 in early stage because force is transmitted from border to center.

Disturbed Stress Field Related to Change in Driving Force
Distribution of $\delta\tau_{max}$ and $\delta\sigma_m$ for model EC24 caused by increase of driving force (by 50%) is shown in Fig.3. It can be seen that the maximum shear stresses in all positions increase with different levels. The higher the essential stress is, the higher the increment is. For example, jog 3 with largest essential stress is of largest incremental stress. Incremental stress within fault zone is 3 orders lower than that in outsides of fault zone. Increment of mean stress is different in different positions. In this model, increment of mean stress shows compressional in right step jog but dilative in left step jog. Three intensively dilative areas and two intensively compressional areas exist. As the orientation of stress trajectories keeps constant, the area extended originally develops further dilation and that compressed originally develops further extrusion. Possible upcoming instability may occur in jog 3 according to maximum distortional strain density criterion[2]. Variation of stress in some positions near jog 3 is very weak, but variation in jog 1 and 2 and other positions may be strong. This indicates that stress disturbance caused by change in driving force is not distributed homogeneously. More strong stress disturbance occurs at positions with original stress concentration.

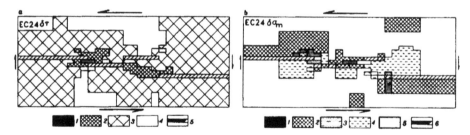

Fig. 3 The disturbed stress field caused by raise of driving force. $\delta\tau$ is increment of maximum shear stress: 1 strongly elevated, 2 elevated, 3 weak elevated, 4 weak variable, 5 faults; $\delta\sigma_m$ is increment of mean stress: 1 strongly dilative, 2 dilative, 3 strongly compressive, 4 compressive, 5 weak variable, 6 faults.

Disturbed Stress Field caused by Redistribution of Stress
Disturbed stress field caused by redistribution of stress for model EC23 and EC24 is studied. In case of existence of elasto-plastic faults, stress level in different positions varies during deformation and sliding along the faults, though the driving force keeps constant. Variations of maximum shear stress and mean stress in some jogs with time are shown in Fig.4. When $t=0.1$, shear stress level in jog 1 and 3 is high, and when $t=0.3$, shear stress level in jog 2 is higher than that in jog 1 and 3. It is clear that at t=0.2, there is a decrease in

maximum shear stress at all positions compared with that at t=0.1. Variation in mean stress is rather strong, and originally dilative area changes to compressive one (denoted by dots) and originally compressive area changes to dilative one (Fig.4 and 5). Strong variation in mean stress occurs in en-echelon jogs.

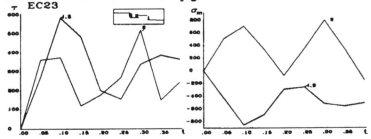

Fig. 4 The stress fluctuation in different positions of model EC23 during the deformation

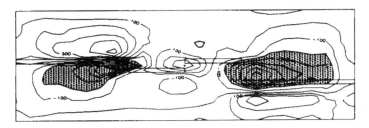

Fig. 5 The increment field of mean stress of model EC24 as comparing the value of t=0.2 to t=0.1. Dotted areas indicate compression increasing area.

Trajectories of stress axes at these three moments are shown in Fig.6. It can be seen that not only stress level but also orientation of stress axes vary with time. For example, orientation of stress axes at t=0.2 are quite different from that at other times.

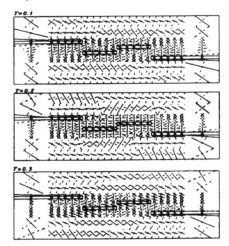

Fig 6 The stress trajectories of model EC23 in different time during the deformation. Solid and dashdot lines indicate orientations of maximum and minimum principal stress, respectively.

Disturbed Stress Field Caused by Fault Propagating and Weakening

In all models, there is a left step en-echelon jog with fault length of 10 and d=1 in center (i.e. jog2). We change elements in this jog from elastic to elasto-plastic and make faults linked to study the disturbed stress field. In this case stress variation transmits from linked area (jog2) to two ends of linked faults (jogs 3 and 1). Areas with rising and descending in maximum shear stress are distributed alternately as an eight-petaline pattern surrounding the linked elements. Areas with descending in τ_{max} are distributed in directions parallel to principal stresses "σ_1" and "σ_2" and areas with rising in τ_{max} are in oblique directions to them (Fig.7a). Dilative and compressive areas are distributed as a four-quadrant pattern surrounding the linked elements in disturbed stress field of mean stress (Fig.7b). Two dilation quadrants align in direction perpendicular to extensional axis and two compressive quadrants in direction perpendicular to compressional axis. Strong descending in τ_{max} and σ_m (that means more compressive) occurs in the propagating and linked elements.

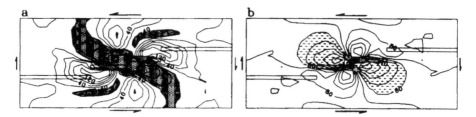

Fig. 7 The disturbed stress field caused by propagation of faults in model EC22. Dotted areas in (a) are areas with descending in maximum shear stress, dashed areas in (b) are areas more compressed.

It is Interesting to indicate that such eight-petaline pattern and four-quadrant pattern of incremental field for maximum shear stress and mean stress have been distinguished not only for en-echelon fault system but also for parallel fault system, bend fault and intersecting fault system. It seems that no matter how complicated the fault system and the stress fields are, the disturbed stress fields always show such patterns as long as the propagation or weakening occurs along a fault. This result provides us a basis to distinguish possible coming unstable fault.

Dynamic Characteristics during Fault Propagating
Preparing process of an earthquake includes propagating and weakening of faults, linking of faults, accelerative slip along fault and so on. All these processes are just prior to instability. Fig.8 shows a disturbed stress field with different dynamic characteristics in a model with left

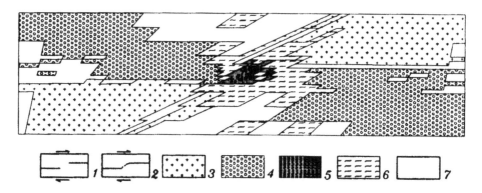

Fig. 8 Dynamic process in mean stress during linking of faults in en-echelon faults. 1 and 2 denote initial and terminal states of faults, 3 continued compressive, 4 continued dilative, 5 strongly undulating, 6 undulating, 7 weak undulating.

step en-echelon faults caused by fault linking under a right lateral shear. The fault propagating begins at the right end of the left fault. The disturbed stress field of maximum shear stress and mean stress keep the same patterns as eight-petaline and four-quadrant patterns, respectively. Interesting to indicate that during the process of fault propagating, there are five kinds of dynamic characteristics in mean stress with time, that is, continued compressive, continued dilative, strongly undulating, undulating and weak undulating process. Continued compressive and dilative types occur roughly in compressive and dilative quadrants, respectively. Undulating type is distributed along the propagating fault and its vicinity. Before Tangshan earthquake in 1976 there was a case history about strongly undulating of water table in a well near the seismogenic fault.

The characteristics in this stage may be summarized as follows: (1) Though essential stress field may be complicated, but disturbed stress fields always keep constant patterns, that is, eight-petaline pattern for $\delta\tau_{max}$ and four-quadrant pattern for $\delta\sigma_m$. Area with strong descending in maximum shear stress and strongly compressive area are located on propagating segment of faults. (2) During fault propagating, changes in stress are very complicated. There are 5 types of dynamic characteristics. Among them, the strongly undulating areas are distributed along the propagating segment and its vicinity. (3) The magnitude of undulation in strongly undulating area can reach that of the essential field. Scale of the strongly undulating area is similar to that of propagating segment, while scale of undulating area is its 3 to 5 times. Stress undulating period coincides with fault propagating period. Its persisting time depends on the propagating rate.

CONCLUSION

Disturbed stress field may be caused by different mechanisms, such as increasing in driving force, stress redistribution, and fault weakening, propagating and linking. Spatial distribution of incremental stress field caused by them is different. This may be the reason for the departure of anomalies in different stages.

The disturbed stress field caused by change in driving force is widely distributed. Orientations of stress axes keep constant and strongly disturbed areas coincide with original areas with stress concentration. However, there is no inevitable relation between strong disturbed area with upcoming hypocenter.

Stress redistribution may change orientation of principal stress axes somewhere and strongly affect mean stress field, specially in en-echelon jog.

Disturbed stress field caused by fault propagating and linking has an eight-petaline pattern for increment of maximum shear stress and a four-quadrant pattern for increment of mean stress.

There are five types of dynamic characteristics in stress disturbance accompanied with fault propagating and linking. Strongly undulating area coincides with propagating segment and their space scales are almost the same.

REFERENCES

1. D.J.Benson and J.Q.Hallquist. A simple rigid body algorithm for structural dynamics programs, *International Journal for Numerical Methods in Engineering*, 22, 723-749(1986).
2. Y. Du and A. Aydin, The maximum distortional strain density criterion for shear fracture propagation with applications to the growth paths of en-echelon faults, *Geophys. Res. Lett.*, 20(11), 1091-11094 (1993).
3. Edition group for "Tangshan earthquake in 1976",*Tangshan earthquake in 1976*, Seismological Press, Beijing, (1982) (in chinese).
4. X. Gao. The precursor process and its field-source features of Datong-Yanggao earthquake, in *"The selected*

papers of earthquake prediction in China", Seismological Press, 195-205 (1996).

5. J. Q. Hallquist. A procedure for the solution of finite deformation contact-impact problems by the finite element method, University of California, *Lawrence Livermore National Laboratory, Rept. UCRL-52066*(1976).

6. J. Q. Hallquist. Theoretical manual for DYNA3D, University of California, *Lawrence Livermore National Laboratory, Rept. UCID-19401*(1983).

7. J. Q. Hallquist, G. L. Goudreau and D. J. Benson. Sliding interfaces with contact-impact in large-scale larrangian   computations, *Comp. mechs. Appl. Mechs. Eng.,* **51,** 107-137(1985).

8. Ma Jin, Y.Du and L.Liu, The instability of en-echelon cracks and its precursors, *J. Phys. Earth,* 34(Suppl), S141-S157 (1986).

9. Ma Zongjin, Fu Zhengxiang, Zhang Yinzhen, Wang Chengming, Zhang Guoming, Liu Defu, Nine strong earthquakes in China from 1966 to 1976, Seismological Press, Beijing, (1982) (in chinese).

10. Mei Shirong et al, *Introduction of earthquake prediction in China,* Seismological Press, (1993) (in chinese).

11. K. Mogi, Fundamental studies in earthquake prediction, *A collection of papers of International Symposium on Continental Seismicity and Earthquake Prediction,* Seismological Press, (1984).

12. Rikitake. *Earthquake prediction,* (chinese version), Seismological Press, Beijing, (1978).

13. G. A. Sobolev. Earthquake prediction in the USSR, fundamental principles, *J. of Earthquake Prediction Research,* 1:1,(1992).

14. M. Wyss. Evaluation of proposed earthquake precursors, *Geophys. Mongr. Am. Geophys. Union,* Washington, D. C., 94, (1991).

15. Zhang Zhaocheng, D. Zheng, Y. Luo and Q. Jia, Studies on earthquake precursors and the comprehensive criteria for earthquake prediction, in *"Continental Earthquakes",* Seismological Press, 215-219 (1993).

Proc. 30ᵗʰ Int'l. Geol. Congr., Vol. 5 pp. 211-221
Ye Hong (Ed)
© VSP 1997

An Experimental Study on Deformation of En-echelon Faults and Seismicity

MA SHENGLI, LIU LIQIANG, MA WENTAO, DENG ZHIHUI, LIU TIANCHANG, AND MA JIN

Institute of Geology, State Seismological Bureau, Beijing 100029, China

Abstract

The samples with compressional, extensional and complex en-echelon faults are deformed experimentally under biaxial compression, spatial and temporal distribution of acoustic emission, fault displacements and strain during deformation are studied. Compressional and extensional en-echelon faults have similar deformation process, i.e. deformation is predominated by fracturing and linking of jogs in the first stage and by sliding along faults in the second stage. However, high strain energy can be accumulated and strong acoustic emission could be produced in compressional jog, while in extensional jog, strong acoustic emission event can not be produced. Deformation process of complex en-echelon faults is not a simple combination of that of two types of en-echelon faults, but contains interaction between them. An interesting phenomenon is that acoustic emission events with high energy occur mostly along main fault near extensional jog, not along fault near compressional jog, but the time of such event is controlled by the deformation process of compressive jog. Instability events with different characteristics could be observed in the experiments. Based on the results, possible relationship between en-echelon faults and seismicity is discussed.

Keywords: rock mechanical test, en-echelon faults, acoustic emission, fault displacement, strain field

INTRODUCTION

It is well known that map traces of strick-slip faults are characteristically discontinuous and en-echelon pattern is the most common fault geometry [6,24,26]. It has been shown that the interaction between en-echelon faults takes important role in the formation and development of both experimental and natural faults [7,25], and high precision microearthquakes studies show that irregularities in surface fault traces may extend throughout the seismogenic zone to depths of 10 km or more [4,20]. Therefore, knowledge of deformation characteristics of en-echelon faults is important to understanding faulting process and related seismicity in the crust.

En-echelon faults can be classified into two types according to property of stress concentration at jog between en-echelon faults, that is, compressional and extensional ones. There has been a lot of studies on the nature of deformation and associated structures at jogs [1,2,7,16]. Stress perturbation fields at jogs and their potential effects on fault behavior have been investigated theoretically by Segall & Pollard[21], Mavko[17], Du et al. [21] and many other authors. Effect of mechanical interaction on propagation and instability for en-echelon faults (cracks) has been suggested by Pollard et al.[18], Pollard & Aydin[19], Sempere & Macdonald[21], Aydin et al.[2] and Ma et al.[13]. Sibson[23] discussed earthquake rupture interaction with fault jogs and suggested that both types of jog may impede earthquake ruptures and lead to their partial or complete arrest. There also have been some experimental studies on deformation and failure process of en-echelon faults [5,9-11,27]. However, most previous analyses were static ones, and most experiments were operated under uniaxial condition and with very limited measurement points, which could not reveal the whole deformation process including fracture and sliding and corresponding physical field. Therefore, further study in deformation and failure process and corresponding physical field is still needed in order to better understand the relationship between en-echelon faults and seismicity.

In this paper, samples with en-echelon faults are deformed experimentally under biaxial compression, and the spatial and temporal evolution of acoustic emission (AE), fault displacement and strain during deformation are measured, and deformation process and instability type and precursors are analyzed. Because we are more interested in effect of instability of en-echelon faults on seismicity, here closely spaced en-echelon faults with strong interaction is considered.

EXPERIMENTAL PROCEDURE

The sample for each experiment is a gabbro block with size of 25cm×25cm×2cm, and sawcuts with   width of   3mm and filled with plaster are made to simulate faults along 45°. One kind of samples contains two non-linear sawcuts in each sample, which forms a set of en-echelon faults. Both distance and overlap distance between two faults are 2cm, and theoretic analysis indicates that there are intense interaction between two faults for en-echelon faults with such geometric texture [8]. The other kind   contains three sawcuts, which forms complex en-echelon faults with two jogs of 2cm×2cm.   Sample textures and sensor locations are shown in Figure 1.

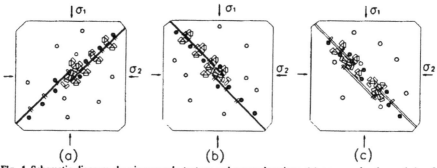

Fig. 1 Schematic diagram showing sample texture and sensor locations. (a) compressional en-echelon faults; (b)extensional en-echelon faults; (c) complex en-echelon faults. Circles denote AE transducers, open and solid ones belong to two AE recording systems, respectively. x-shaped symbols denote fault displacement gauges, and rectangle symbols denote strain gauges.

The tests are operated on a biaxial rig, and all samples are deformed at lateral stress (σ_2) of 5 MPa and axial shortening rate of 0.5 μm/s. Axial and lateral stress are applied simultaneously to 5 MPa first, and then make sample deformed at the constant axial shortening rate.

In the experiments, besides measuring the axial loading stress and displacement, AE transducers, strain gauges, fault displacement gauges are attached on samples to measure the corresponding physical fields. Waveforms of AE events are recorded by two digital systems with high sampling rate and multi-channels. The location of an AE event can be determined by its arriving time to different transducers. Its magnitude is determined by using the method in seismology and considering the difference between AE events and earthquakes [14]. In this way, M-t diagram, spatial and temporal distribution of AE events during deformation for each sample can be obtained. In this paper, we only analyze AE data from one recording system (its transducers denoted by solid circles in Fig.1). The strain is recorded by a digital system with low sampling frequency, high resolution and multi-channels. In the experiments, 32 standard strain gauges are used for strain measurement. The fault displacement distribution is measured by the same system with 8 special gauges. Details of the experimental system have been described by Liu & Liu[12].

RESULTS

Figure 2 shows secondary fractures developed during deformation for samples with different types of en-echelon faults. Two shear fractures occur in compressional jog(Fig.2a), and four tensile fractures in extensional jog (Fig.2b). Fracture pattern is more complicated in complex en-echelon faults, but basically it is like a combination of that in compressional and extensional en-echelon faults except more cracks in wider area near the compressional jog (Fig.2c). These fractures take important role in controlling evolution of physical field.

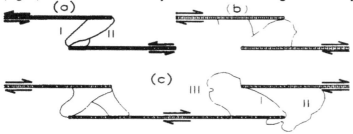

Fig. 2 Sketches showing fractures in jogs. (a) compressional en-echelon faults; (b)extensional en-echelon faults; (c) complex en-echelon faults.

Compressional en-echelon faults
Figure 3 shows differential stress-time curve for sample with compressional en-echelon faults. Stress increases steadily with time (strain hardening). Two obvious stress drops

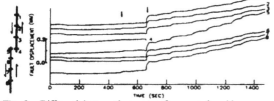

Fig. 3 Differential stress-time curve for sample with compressional en-echelon faults. Arrows indicate instability events.

(instability events) occur at ~450s and ~650s, respectively, and only small vibrations in stress occur periodically after ~650s.

Fault displacement shows obvious changes with time and space (Fig. 4). It increases with time very slowly before ~650s but quickly after

~ 650s, and displcement at the ends of faults is greater. There are only small step changes in fault displacement at some parts of faults associated with the first stress drop, but big step changes along the whole faults corresponding to the second one. This indicates that there may be a transition in predominant deformation mode from fracturing to sliding at ~650s, and instability events at ~450s and ~650s may be caused by fracturing and unstable sliding,

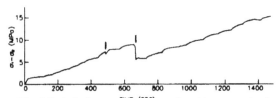

Fig. 4 Fault displacement-time curves for with compressional en-echelon faults. Arrows indicate instability events

respectively.

Figure 5 shows strain-time curves for strain gauges parallel to fault strike and in the jog. Here positive value means compression and negative means extension. There is no big difference in initial strain (strain just before applying differential stress) along faults except that the inner ends of faults (No.12 and 31) have higher value. During deformation, strain in areas far from the jog has little change (No.3 and 19), while strain in the jog and its vicinity changes greatly and complexly. Dramatic drop of strain in gauge No.15 at ~100s indicates the starting of crack I (see Fig. 2a). Crack II (see Fig.2a) occurs and propagates at ~220s, which makes strain in gauge No.16 change abruptly. As deformation continues, crack I propagates slowly, causing strain to decrease steadily along the crack (No.15 and 28) and to increase in adjacent area (No.29, 12, 31). The crack is linked at ~450s, leading to a rapid release of strain in area near it and an obvious increase of strain in adjacent area. Crack II propagates acceleratively at ~580s, causing an great increase of strain in gauge No.29. The crack is linked at ~650s, causing a strongly rapid release of strain along it and at the inner ends of the faults. After ~650s, strain increases in the jog and along inside of faults but decreases along outside of faults.

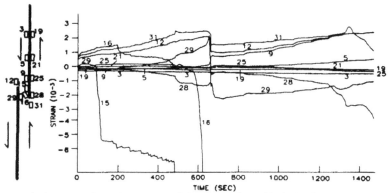

Fig. 5   Strain-time curves for strain gauges parallel to fault strike and in the jog for sample with compressional en-echelon faults.

Figure 6 shows spatial and temporal distribution of AE events for sample with compressional en-echelon faults. It can be seen   that AE activity   exhibits inhomogeneous distribution with time (Fig.6a). AE produce frequency is low before 400s, and there are two active periods in AE corresponding to stress drops at ~450s and ~650s respectively. After that, AE produce frequency lowers again, but it seems that AE activity becomes active with progression of deformation. In space, almost all AE events occur   near faults. In the first stage (before ~650s), AE events are distributed mainly near the jog,   and have tendency to spread from the jog to faults. In the second stage, most AE events occur along faults off the jog (Fig.6b).

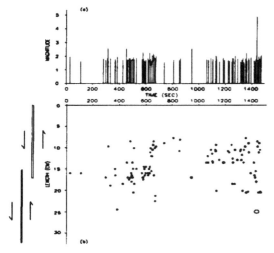

Fig.6 M-t diagram(a) and spatial-temporal distribution (b) of AE events for sample with compressional en-echelon faults

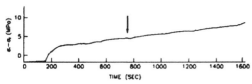

Fig. 7 Differential stress-time curves for sample with exten-sional en-echelon faults. Arrow indicates the main instability event.

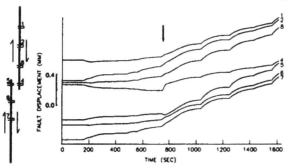

Fig. 8 Fault displacement-time curves for sample with extensional en-echelon faults. Arrow indicates instability event.

Extensional en-echelon faults

Compared with sample with compressional en-echelon faults, sample with extensional en-echelon faults is weaker and strain hardening is also weaker. There is a stress drop at about 750s and small stress drops with period of 200~300s occur in succeeding deformation (Fig.7).

Fault displacement along extensional faults (Fig. 8) shows obvious changes with time and space, and is more complicated than that along compressional faults. Before ~750s, fault displacement increases slowly with time, and faults near the jog almost do not slip even slip backward, while middle parts of faults slip quickly. After ~750s, displacement along faults increases greatly with time. There are obvious changes in fault displacement along the whole fault systems corresponding to the stress drop at ~750s. This shows that the transision of deformation model from fracturing to sliding occurs at ~750s.

Figure 9 shows strain-time curves for strain gauges parallel to fault strike. It can be seen that initial strain are extensional within the jog and along faults but compressional at the inner ends of faults. After differential stress is applied, strain gauges (No.15 and 29) in the jog are broken successively, indicating the occur-rence of cracks in the jog (see Fig.2b) and causing strain to decrease slightly near the jog and along faults but to increase strongly near the inner ends of faults and at outside of the jog. After that, strain changes gently, and there is a series of small release of strain occurring periodically along inside or outside of faults. Corresponding to the stress drop at ~750s, there is a process with strong release of strain, mainly along outsides of the jog (No.31 and 12) and of faults (No.25 and 28). This process indicates the complete linkage of the jog. After this event, there are periodic changes in strain, mainly along middle part of fault.

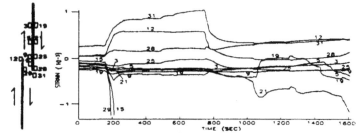

Fig. 9   Strain-time curves for strain gauges parallel to fault strike for sample with extensional en-echelon faults.

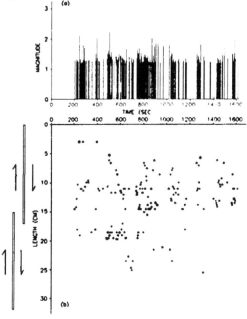

Corrpared with compressional en-echelon faults, AE events of sample with exten-sional en-echelon faults are smaller in magnitude (Fig.10a). AE activity exhibits obvious periodicity, especially after ~800s, which corresponds to stress vibrations in time, and there is a strong period in AE corresponding to the stress drop at ~750s (Fig. 10a). In space, AE events are distributed mainly along faults near the jog but seldom within the jog (Fig.10.b). It should be pointed, however, that more AE events are recorded by the other AE recording system (its transducers denoted by open circles in Fig.1) with lower triggering level, and most of them are distributed within the jog [15]. This indicates that extensional jog may only produce very small events, and bigger events are located at faults off the jog.

Fig 10 M-t diagram (a) and spatial-temporal distribution (b) of AE events for sample with extensional en-echelon faults.

Complex en-echelon faults

Sample with complex en-echelon faults is as strong as sample with compressional en-echelon faults. Strain hardening is rather strong before ~350s but becomes weaker after ~350s. A big stress drop occurs at about 350s and small stress drops occur periodically in succeeding deformation (Fig.11).

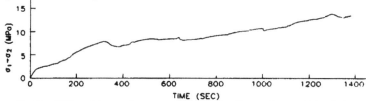

Fig. 11   Differential stress-time curves for sample with complex en-echelon faults.

Changes in strain are rather complicated, revealing the complex deformation process (Fig.

12). Because of heterogeneity in sample texture, there exists difference in initial strain. The general characteristics are that strain is negative in extensional jog and positive in compressional jog. At the first stage of deformation, deformation is mainly concentrated on vicinity of extensional jog. When differential stress is applied, cracks in the jog (see Fig.2c) occur and propagate successively, indicated by breaking of strain gouges No.7 and 8 (see arrows ①② in Fig.12), which is accompanied by strain release mainly near the jog (No.19 and 5). At ~350s, there is a process with strong release of strain near extensional jog and along outside of the main fault, corresponding to the complete linkage of the jog (arrow ④ in Fig.12). Deformation of extensional jog is accompanied by deformation of compressional jog, presented by starting of crack I in compressional jog (see Fig.2c)

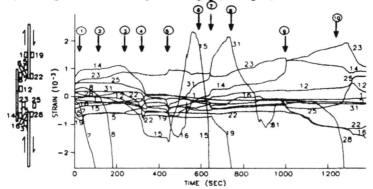

Fig. 12 Strain-time curves for strain gauges parallel to fault strike and in jogs for sample with complex en-echelon faults. ①② occurrence and propagation of cracks in extensional jog; ③ occurrence of crack I in compressional jog; ④ propagation of crack I in compressional jog and linkage of extensional jog; ⑤ occurrence of crack II in compressional jog; ⑥ linkage of crack I in compressional jog; ⑦ accelerative slip of the main fault in extensional jog; ⑧ propagation and linkage of crack II in compressional jog; ⑨ occurrence of crack III in compressional jog; ⑩ propagation and linkage of crack III in compressional jog.

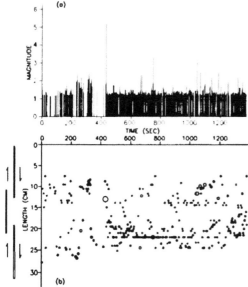

Fig. 13 M-t diagram (a) and spatial-temporal distribu-tion (b) of AE events for sample with complex en-echelon faults.

at time indicated by arrow ③ in Fig.12 and its accelerative propagation at time when extensional jog is linked (gauges No.14 and 15). After ~400s, deformation is concentrated on vicinity of compressional jog, which is shown by occurrence and propagation of crack II and III and linkage of crack I in the jog (see Fig.2c), causing complicated process of strain accumulation and release (see arrows and explanation in Fig.12). At ~620s (arrow ⑥ in Fig.12), for example, linkage of crack I in compressional jog causes a process with strong release of strain, leading to the breaking of some strain gauges (No.15 etc.) and triggering a rapid slip with strong release of strain along the main fault near extensional jog (arrow ⑦ in Fig.12). After this event, along with the further deformation of compressional jog, accelerative slip occurs along the main fault in extensional jog periodically, accompanied by strain release (gauges No. 19, 22, etc.)

AE activity is of ralatively low produce frequency and small difference in magnitude before ~400s, but there is a strong AE sequence corresponding to stress drop at ~350s. After ~420s, AE activity exhibits very high produce frequency and big difference in magnitude, and big events occur periodically, corresponding to stress drops in time (Fig.13a). In space, AE events are distributed mainly at two jogs and along nearby faults. AE activity near extensional jog is of obvious periodicity, and most strong AE events are distributed along the main fault near the jog. AE activity near compressional jog is of very high produce frequency after ~420s, but strong AE events are not common (Fig.13b).

DISCUSSIONS

Deformation processes of en-echelon faults
Compressional and extensional en-echelon faults have similar deformation process. The deformation is   predominated by fracturing and linking of the jog in the first stage, strain changes intensely but fault slip is quite small, and AE activity has high produce frequency but low energy (small magnitude). The deformation is predominated by sliding along faults in the second stage, fault slip is large, and strain changes relatively small.   In compressional jog, however, high strain energy can be accumulated and intense release of strain and AE event can be produced, and the jog has a permanent resistance to fault sliding. While in extensional jog, rapid release of strain and strong AE event can not be produced, and the jog has almost no resistance to fault sliding in the second stage. In general, the results are similar to previous studies[8,11,13], showing that AE magnitude is related with stress level.

The deformation   process of complex en-echelon faults is not a simple combination of that of two types of en-echelon faults, but contains interaction between them. An interesting phenomenon is that although the strength and maximum magnitude of AE are similar to compressional en-echelon faults, bigger AE events with high energy occur mostly along the main fault near extensional jog with lower stress level, not along the faults near compressional jog with higher stress level,   but the time of such events is obviously controlled by the deformation process of compressional jog. The results are quite different from previous studies and indicate that AE magnitude is not only related with stress level, rapid sliding along fault   under some conditions may also produce strong AE events.

Instability type and precursor characteristics along en-echelon faults
Instability events with different characteristics are observed in the experiments, which can be divided into fracturing type, stick-slip type and mixed type based on their deformation mechanism and evolution characteristics of physical field. Different type instability differs greatly in precursor characteristics. Here we choose some typical events to show characteristics of each type instability and precursors.

Fracturing type instability can be caused by fractures in compressional jog. Figure 14 shows variations in measuring parameters before and after such an instability event caused by rapid propagation of crack I (see Fig.2a) in compressional en-echelon faults. Total stress begins to increase tens of seconds prior to instability. There are rapid changes in fault displacement associated with the instability in some parts of faults, but change before the instability is not obvious. The sequence of AE events indicates that there are a few precursory events but a lot of events similar to aftershocks. The strain release associated with instability is mainly concentrated on the crack and outside of faults, and the strain decreases clearly near the crack but increases in area around it before the instability.

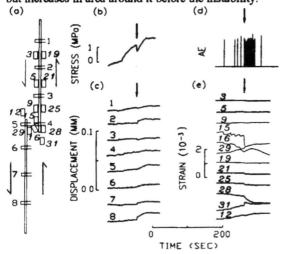

Fig. 14 Changes in various parameters before and after an instability event during fracturing compressional jog. (a) fault texture and distribution of transducers; (b) stress; (c) fault displacement; (d) acoustic emission; (e) strain.

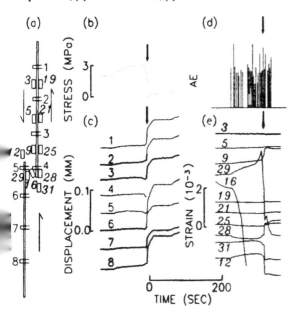

Fig.15 Changes in various parameters before and after an instability event caused by linking of compressional jog. (a) fault texture and distribution of transducers; (b) stress; (c) fault displacement; (d) acoustic emission; (e) strain.

Mixed type instability involves both fracturing and sliding, and may occur when en-echelon faults are linked. Propagation of secondary fracture makes jog linked, which leads rapid slip along whole fault system and cause instability. Figure 15 shows variations in measuring parameters before and after an instability event caused by linkage of the jog along compressional en-echelon faults. Total stress begins to increase about 100s before the instability and decreases just before the instability. Fault segments far-off the jog slip acceleratively, while fault segments near the jog slip backward first and then acceleratively before the instability. Precursor phenomenon in AE is obvious. From the result of strain measurement, it appears that the instability occurs after the linking of crack II (see Fig.2a), and the strain release is concentrated on the lower part of the jog (No.29) and its outside (No.12 and 31). Strain in these positions, especially in position No.29, increases acceleratively before the instability and then decreases rapidly just before the instability, while strain in other areas shows opposite tendency.

Stick-slip type instability occurs during sliding after linkage of jogs along en-echelon faults. Because of low lateral stress in our experiments, such type instability is small and we will not give special analysis. What warrants to explain is that some stick-slip events have obvious precursors and some have not, depending on if there is strain transfer between different parts of fault.

This is similar to that in typical rock friction tests and may be related with stick-slip mechanism [14].

En-echelon faults and seismicity

Here we make some discussions on deformation of en-echelon faults and seismicity based on our experimental results.

In the jog of compressional en-echelon faults, subfracturing may cause moderate earthquakes, and strain field may have precursory changes before such an earthquake. In the jog of extensional en-echelon faults, subfracturing may only cause small earthquakes. Linking of jogs along en-echelon faults may cause strong earthquakes, and such an earthquake may have very rich precursors in strain field, fault displacement and seismicity. For fault system with compressional and extensional jogs, compressional jog may play important role in controlling deformation along the whole fault system. Deformation and failure of such area can provide yielding condition necessary for slip instability along faults, so an seismo-active period along a fault system may begin from such areas. Strong earthquakes may occur along main faults near both of extensional and compressional en-echelon jogs.

REFERENCES

1. A. Aydin and A. Nur. Evolution of pull-apart basins and their scale independence, *Tectonics* 1, 91-105(1982).
2. A. Aydin and A. Nur. The types and role os stepovers in strick-slip tectonics. In: *Strick-slip Deformation, Basin Formation, and Sedimentation.* K.T. Biddle and N. Christie-Blick (Eds.). Spec. Publis. Soc. econ. Paleont. Miner. 37, 35-44 (1985).
3. A. Aydin and R. A. Schultz. Effect of mechanical interaction on the development of strick-slip faults with echelon patterns, *J. Struct. Geol.* 12, 123-129 (1990).
4. W.H. Bakun, R.M. Stewart, C.G. Bufe and S.M. Marks. Implication of seismicity for failure of a section of the   San Andreas fault, *Bull. Seism. Soc. Amer.* 70, 185-201 (1980).
5. E.G. Bomblakis. Study of the brittle fracture process under uniaxial compression, *Tectonophysics* 18, 261-270 (1973).
6. M.G. Bonilla. Historic faulting-map patterns, relation to surface faulting and relation to pre-existing faults, *U.S.G.S. Open File Report* 79-1239, 36-65(1979).
7. Q. Deng and P. Zhang. Research on the geometry of shear fracture zones, *J. Geophys. Res.* 89, 5699-5710 (1984).
8. Y. Du, J. Ma and J. Li. Interaction and stability of en-echelon cracks, *Acta Geophys. Sinica (in Chinese)* 32, Suppl . (I), 218-231 (1989).
9. J. Li, X. Wu, B. Zhang and T. Liu. Experimental studies on failure processes of saw-cut rocks under confining pressure, *Seismology and Geology* 6(2), 76-80 (1984) (in Chinese).
10. J. Li, G. Shi and J. Ma. Study of fractures and instability pattern in the region of en-echelon cracks, *Research on Recent Crustal Movement* 4, 149-155 (1989) (in Chinese).
11. L. Liu, J. Ma and X.Wu. An experimental study on the process of deformation and instability for en-echelon faults, *Acta Seism. Sinica* 8, 393-403 (1986) (in Chinese).
12. L. Liu and T. Liu. Design and trial-manufacture of the system for measuring physical fields of tectonic deformation in laboratory, *Seismology and Geology* 17, 357-362 (1995) (in Chinese).
13. J. Ma, Y. Du and L. Liu. The instability of en-echelon cracks and its precursors, *J. Phys. Earth* 34, Suppl., 141-157 (1986).
14. J. Ma, S. Ma, L. Liu, Z. Deng, W. Ma and T. Liu. Geometrical textures of faults, evolution of physical field and insta-bility characteristics, *Acta Seism. Sinica* 9, 261-269 (1996).
15. W. Ma, J. Ma, L. Liu, S. Ma, T. Liu and   Z. Deng. The characteristics of acoustic   emission in   en-echelon structure, *Seismology and Geology* 17, 342-348 (1995) (in Chinese).
16. P. Mann, M.R. Hempton, D.C. Bradley and K. Burke. Development of pull-apart basin, *J. Geol.* 91, 529-554 (1983).
17. G.M. Mavko. Fault interaction near Hollister, California. *J. Geophys. Res.* 87, 7,807-7,816(1982).
18. D.D. Pollard, P. Segal and P.T. Delaney. Formation and interpretation of dilatant echelon cracks, *Bull. Geol. Soc. Am.* 93, 1291-1303(1982).

19. D.D. Pollard and A. Aydin. Propagation and linkage of oceanic ridge segments, *J. Geophys. Res.* **89**, 10,017-100,28(1984).

20. P. Reasenberg and W.L Ellsworth. Aftershocks of the Coyote Lake, California, earthquake of August 6, 1979: a de-tailed study, *J. Geophys. Res.* **87**, 10,637-10,655 (1982).

21. P. Segall and D. Polland. Mechanics of discontinous faults, *J. Geophys. Res.* **85**, 4337-4350 (1980).

22. R. H. Sibson. Rupture interaction with fault jogs, *Amer. Geophys. Un. Geophys. Mon.* **37**, 157-368 (1986).

23. J.-C. Sempere and K.C. Macdonald. Overlapping spreading centers: implications from crack growth simulation by the displacement discontinuity method, *Tectonics* **5**, 151-163(1986).

24. R.V. Sharp. The implication of surfacial strick-slip fault patterns for simplification and widening with depth, *U.S.G.S. Open File Rep.* **79-1239**, 66-78(1979).

25. J.S. Tchalenko. Similarities between shear zones of different magnitudes, *Bull. geol. Soc. Am.* **81**, 1625-1640(1970).

26. R.E. Wallace. Surface fracture patterns along the San Andreas Fault. In: Proc. *Conf. on Tectonic Problems of the San Andreas fault system.* R.L. Kovach and A. Nur (Eds.). pp.248-250. Stanford (1973).

27. Z. Zhang and Q. Li. Experimental Study of the propagation process of fracture system and variation characteristics of geophysical field, *Progress in Geophysics* **8**(4), 225-231(1993) (in Chinese).

Proc. 30ᵗʰ Int'l. Geol. Congr., Vol. 5 pp. 223-232
Ye Hong (Ed)
© VSP 1997

Temperature Measurements along Simulated Faults during Seismic Fault Motion

AKITO TSUTSUMI AND TOSHIHIKO SHIMAMOTO

Earthquake Research Institute, The University of Tokyo, 1-1-1 Yayoi, Bunkyo-ku, Tokyo 113, Japan

Abstract

A series of high-velocity frictional experiments on monzodiorite and granite at high velocities (to 13 m/sec), large displacements (97-1433 m) and low normal stresses (to 1.5 MPa) are conducted to measure the temperature rise and friction along simulated faults. Temperature rise is measured with PR and CA thermocouples and with a radiation thermometer. Measured average temperature along simulated faults during frictional melting reaches about 1070 -1190 °C for monzodiorite and 1130 -1145 °C for granite, well above the decomposition temperature of biotite (650 °C). This temperature is consistent with an almost complete lack of biotite clasts in experimentally produced pseudotachylyte.Thus the frictional melting is a selective melting process. Measured temperature in monzodiorite specimens agrees with calculated temperature distribution by use of the finite element method.

Initial strength peak in friction is followed by a gradual decay in friction with increasing displacement upon the initiation of experiment. Friction, however, begins to increase as the fault is heated, undergoes a broad peak and eventually approaches a steady-state. The moment when the simulated fault attains steady-state temperature distribution corresponds to the moment when the fault surface reaches steady-state mechanically. Visible frictional melting begins to occur near the second broad peak in friction, hence well before the steady-state temperature distribution is attained.

Keywords: rock friction, pseudotachylyte, frictional melting, frictional heating

INTRODUCTION

Occurrences of pseudotachylyte along exhumed faults suggest that seismic fault motion produces sufficient heat to cause frictional melting [e.g., 20]. Tsutsumi and Shimamoto [24] have shown recently that frictional heating and frictional melting changes the frictional properties of simulated faults dramatically, as suggested by Mckenzie and Brune [11]. Mechanical properties of rocks are so sensitive to temperature that the evaluation of temperature rise along a fault is essential to establish the fault constitutive properties at high velocities and to understand the underlying physical processes.

Some workers measured temperature rise along simulated faults in laboratory for stable sliding and for the stick-slip mode sliding [8, 22, 26, 27]. The slip rate during stick-slip events reached several tens of cm/sec, a seismogenic velocity [6, 16]. However, the displacements during stick-slip were not sufficient to cause bulk frictional melting along simulated faults. We report herein results from our temperature measurement along simulated faults at high-velocities and under large displacements, favorable for frictional melting.

APPARATUS AND EXPERIMENTAL PROCEDURES

All experiments were performed using a rotary-shear, high-speed frictional testing machine at our Institute [18], which is capable of producing slip rates to 1.3 m/sec for a cylindrical specimen of 25mm in diameter, an unlimited displacement, and axial forces to 10 kN with an air-driven actuator (normal stresses to 20 MPa for our specimens). One side of the specimens is kept stationary during experiments, while the other side is rotated directly with a 7.5 kw servomotor with; the maximum revolution rate to 1500 r.p.m. For cylindrical specimens, the slip rate is zero in the center and at maximum at the circumference. We use the equivalent slip rate, Y_e, defined such that V_e multiplied by the area of fault surface gives the rate of frictional work assuming constant frictional coefficient over the fault [18].

Rock sample used in the experiments are a fine-grained monzodiorite (of unknown origin) and a fine-grained Inada granite from Tsukuba, northeast of Tokyo. All samples are equigranular and have no planer structures. Monzodiorite consists of k-feldspar, plagioclase, pyroxene, biotite and magnetite and the average grain diameter is about 0.3 mm. Inada granite consists of quartz, plagioclase, k-feldspar, hornblende, muscovite and biotite and the average grain diameter is about 1.5 mm. Simulated faults, i.e., end-surfaces of solid cylindrical specimens, were ground with 100# grid silicon-carbide grinding wheels.

Uniaxial strength of specimens decreases by more than two order of magnitude due to thermal fracturing associated with the frictional heating [12]. Experiments were thus performed at low normal stresses (a,) to 1.5 MPa, with more than 97 m of total displacement. Displacements may appear unrealistically large for simulating seismic fault motion. In terms of frictional work, however, a displacement of 100 m at σ_n =1.5 MPa corresponds to 1 m displacement at sn=150 MPa, assuming the same frictional coefficient. Thus our experiments are not necessarily unrealistic for simulating fault motion at depth.

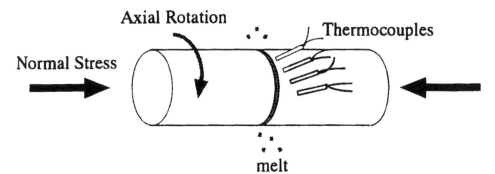

Fig.1   A schematic sketch showing the specimen assembly for frictional melting experiment.

Thermocouples (CA or PR) and a radiation thermometer (MINOLTA, TR-630A) were used for temperature measurements. PR- and CA-thermocouples of 0.25 mm and 0.1 mm in diameter, respectively, were set in 1 to 4 holes drilled at an angle of 40°- 60°to the long axis of the specimen of stationary side (Fig.1). The diameters of the holes are, respectively, 1.5 mm for the PR thermocouple and 1 mm for the CA-thermocouple. Thermocouples were buried in the hole with alumina powder and were cemented with heat-resisting cement at the mouth (THERMON T-63 of Hakko-shouji Co.Ltd. and Ceramabond 569 of Amerco Co. Ltd.). The length of specimens decreases with increasing displacement as the gouge and/or frictional melt drop out of the simulated fault. Hence, the distance between the thermocouple and the sliding surface decreases during a run, and the thermocouple fails soon after it meets the sliding surface. The thermocouple would touch and detect the temperature of the melt product at the moment immediately prior to the thermocouple failure. The thermocouple closest to the sliding surface was placed 0.5 to 2 mm from it so that the thermocouple would meet the surface in a few minutes. Since the slip rate on the sliding surface increases outwards from the center, the rate of heat production is at maximum at the circumferential part. On the other band, the heat is lost from the surface of the specimen, so that the maximum temperature is expected somewhere between the circumference and the center. Thus the thermocouple were placed at around 2 to 5 mm from the periphery. This location is close to the location of the maximum temperature asshown later by numerical simulation.

A radiation thermometer is a device to determine the temperature of the surface of the object by measuring the radiation energy emitted from the surface. The thermometer used in this study can measure the average temperature of a surface as small as 0.4 mm2 from the distance of 0.2 m, for the temperature range of 550 °C to 3000 °C. Test condition and experimental results are summarized in Table 1.

Table 1 Summary of data from the frictional melting experiments.※, Thermocouple could not reach to the sliding surface during a run: di, monzodiorite: gr, Inada granite: PR, PR-thermocouple: CA, CA-thermocouple: Rd-Therm, Radiation thermometer: s normal stress. Slip rate is the equivalent velocity, V_e, and total displacement refers to the displacement at the outer edge of the cylindrical specimen.

Run number and Sample	σ (MPa)	Slip Rate (m/s)	Total Displacement (m)	Temp.-measuring Device	Measured Maximum Temp. (°C)
HFR057 di	0.4	0.87	1433	CA	760 *
HFR063 di	1.0	1.3	270	PR	1090
HFR064 di	1.0	1.3	226	PR	1070
HFR065 di	0.8	1.3	650	PR	950 *
HFR066 di	0.9	1.3	135	PR	1100
HFR067 di	0.9	1.3	320	PR	1124
HFR068 di	0.8	1.3	192	PR	1102
HFR069 di	0.8	1.3	257	CA	1085
HFR070 di	0.8	1.3	574	PR	1100
HFR074 di	1.0	1.3	530	Rd-Therm	1169
HFR075 di	1.3	1.3	522	Rd-Therm	1082
HFR076 gr	0.8	1.3	97	Rd-Therm	1145
HFR077 gr	1.0	1.3	102	Rd-Therm	1130
HFR078 di	1.0	0.88	654	Rd-Therm	1090
HFR135 di	1.5	1	216	CA, Rd-Therm	1090(CA), 1190(Rd-Therm)

EXPERIMENTAL RESULTS

Temperature measurement was made with thermocouples or with a radiation thermometer for monzodiorite, whereas only a radiation thermometer was used for granite since the use of thermocouple was quite difficult due to severe thermal fracturing of quartz-containing rocks (Table 1). A typical result of the temperature measurement for monzodiorite, using both four CA thermocouples and the radiation thermometer simultaneously is shown in Fig 2b, in

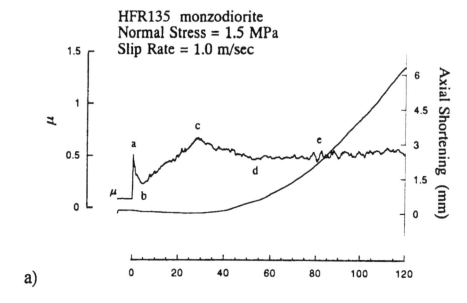

a)

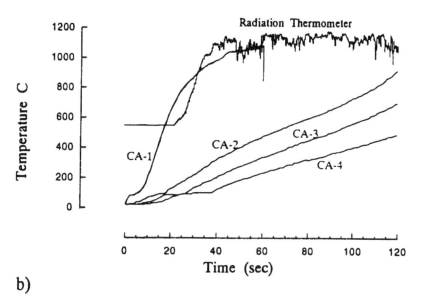

b)

Fig. 2 (a) Frictional coefficient (μ) versus time and axial shortening versus time records from a high-velocity frictional experiment on monzodiorite (HFR135). **(b)** Plot of temperature rise as measured by four CA-thermocouples and a radiation thermometer against elapsed time in the experiment shown in (a). Initial distances of the thermocouples CA-1, CA-2, CA-3 and CA-4 are 0.4, 5.3, 6.8 and 9.4 mm, respectively. "fm" denotes the moment when visible frictional melting initiated.

which temperature is expressed in terms of degrees centigrade against the elapsed time. Changes in the coefficient of friction and axial shortening of the specimen during the run is also shown in the same figure for comparison (Fig.2a).

The initial strength peak (a in Fig. 2a) is followed by a gradual decay in friction with increasing displacement upon the initiation of experiment. Friction, however, begins to increase as the fault is heated (b to c in Fig.2a), undergoes a broad peak and eventually approaches a steady- state (near e in Fig. 2a). Visible frictional melting initiated near this peak (c in Fig. 2a). The rate of shortening is nearly constant when the steady-state friction is attained (Fig. 2a).

After the visible frictional melting begins to occur at about 28 seconds (Fig.2a), the distance between the sliding surface and the thermocouple decreases. The measured temperature by the nearest thermocouple to the sliding surface (CA-1) initially increases rapidly and then gradually (Fig.2b). The CA-1 thermocouple meets the sliding surface and fails at the elapsed time of about 59 seconds(total circumferential displacement=91.7m). The maximum recorded temperature measured by this thermocouple is 1090°C. Other thermocouples do not reach the sliding surface during this run. The measured temperature using the radiation thermometer initially increases rapidly and then reaches to a nearly constant temperature of about 1150°C at the elapsed time of about 43 seconds (Fig.2b, total circumferential displacement=66.8 m). Large variation of the temperature measured by the radiation thermometer is due to defocusing of the sliding surface, since the thermometer was held with our hands. The maximum temperature measured by this device is 1190°C. Thus the measured temperature values with a radiation thermometer are in reasonable agreement with those measured with thermocouples (Table 1).

Fig. 3 shows results of temperature measurements for granite using the radiation thermometer, the maximum being 1145 °C.

DISCUSSION AND CONCLUSIONS

Recent chemical and petrographical studies of the composition of pseudotachylytes have shown that selective decomposition of hydrated ferromagnesian minerals into the frictional melt is responsible for the composition of the final pseudotachylyte products [1,7, 9, 21]. In this context the classical hypothesis suggesting that pseudotachylyte melt forms by equilibrium partial or equilibrium total melting of host rock seems no longer be appropriate for the process of the frictional melting. Thus the estimation of temperature rise during pseudotachylyte formation based on the assumption that the melting of minerals was controlled by equilibrium melting is invalid.

The measured temperature at the simulated faults during the frictional melting reached 1070°C ~1190°C for monzodiorite and to 1130~1145°C for granite (Table 1). These results are consistent with nearly complete lack of biotite in the glassy matrix [23], for the temperature is well above the decomposition temperature of biotite under atmospheric pressure (650°C, [3]). Plagioclase and clinopyroxene clasts in the glassy matrix of monzodiorite are subangular to round-shaped (Fig. 2b of Tsutsumi and Shimamoto [24]), indicating dissolution of this mineral in the melt. Since the melting temperature of plagioclase and clinopyroxene contained in the monzodiorite (andesine and augite) is about 1200°C and 1280°C [14] , respectively, at the atmospheric pressure, the real temperature in the melt must have exceeded 1280°C at least locally. Furthermore, the composition of glassy matrix of granite suggests melting of dry quartz [17] and so the temperature must have exceeded locally even the melting temperature of quartz (1713°C). Clasts of magnetite whose melting temperature is 1590 °C are mostly angular in shape, so that the temperature in the major part of the melt must have been below this melting temperature. These results suggest (1) heterogeneous temperature distribution within a fault zone and (2) a selective melting process. This is not surprising since now we recognize that frictional melting is a rapid, non-equilibrium process [e.g., 17].

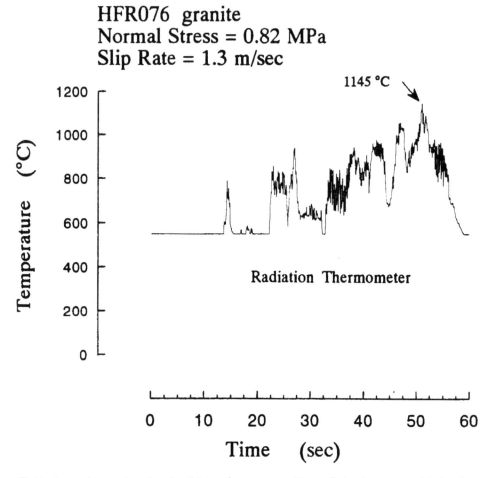

Fig.3 Averaged temperature along the sliding surface as measured by a radiation thermometer, plotted against the elapsed time for Inada granite (HFRO76). Large variation of the measured temperature is due to defocusing of the thermometer on the sliding surface, since we hold the thermometer with hands during the experiment.

We have conducted numerical analysis of temperature distribution within the specimen in order to test whether the observed temperature can be predicted from the analysis (Fig. 4). The calculation was made as a linear axi-symmetric problem using a published computer program of the finite element method after Shimoseki and Fujinuma [19]. It is assumed that all the frictional work is dissipated as heat and that the cooling from the cylindrical surface of specimen obeys Newton's law of cooling [2]. Heat loss due to the continuous drop of

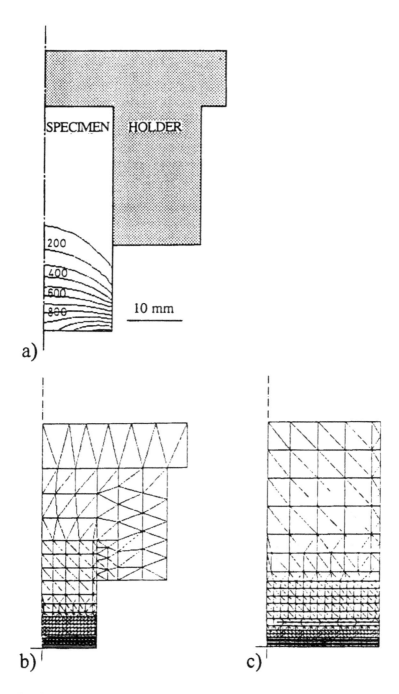

Fig.4 (a) The calculated steady state temperature distribution of the specimen under the experimental condition for frictional melting of monzodiorite (HFRI35). Only a half of the specimen and the sample holder is illustrated here. Temperature is calculated as an axisymmetric problem by using a computer program of finite element method. Intervals of the isothermal curves are 100℃. (b) Mesh used in the calculation used in this study. 494 nods and 890 elements. (c) Enlarged figure of (b) for the specimen.

frictional melt from the sliding surface was also incorporated in the calculation. For the calculation: m (coefficient of friction)=0.48 (steady-state friction in Fig. 2a); k (thermal conductivity)=0.007 cal/cm s °C for specimen and 0.12 cal/cm s°C for specimen holder [3] ; R (revolution rate)=20 r.p.s; sn (normal stress)=1.5 MPa and h (coefficient of heat transfer for flowing air)=0.002 cal/cm2s°C.

Fig.4 displays the calculated steady-state temperature distribution of the specimen; the temperature of the simulated fault surface is the highest at around 2 mm from the periphery and decreases inwards and outwards. The calculated temperature was about 1260°C at maximum and about 1200°C at the circumferential part of the simulated fault. The latter is in reasonable agreement with the measured maximum temperature using a radiation thermometer (Fig.2b).

The measured temperature in Fig.2 is plotted in Fig.5 against the distance of thermocouples from the fault surface along with the calculated temperature for steady-state. To evaluate the change in the distance of thermocouples to the sliding surface during a run, we have made following assumption: each half of the sample erodes at the rate calculated from the final shortening ratio of each half of sample. For example, at the end of the experiment shown in Fig. 2, final shortening of the stationary side specimen was 60 % of the total shortening. So the decrease in depth of each thermocouple was calculated assuming that the stationary side of the specimen eroded at this rate. The calculated steady-state temperature distribution is consistent with the measured temperature distribution of the specimen after the elapsed time of about 81 seconds(Fig. 5), although the temperature is somewhat overestimated in the analysis. Results shown in Fig. 5 indicates that the fault surface attained steady-state temperature distribution at the elapsed time of 81 seconds. This moment when the simulated fault attains steady-state temperature distribution corresponds to the moment when the fault surface reaches steady-state mechanically (Fig. 5).

Visible frictional melting begins to occur near the second broad peak in friction, well before the steady-state is attained. Our temperature calculation is based upon a constant frictional coefficient, and changing friction during a run is not incorporated in the analysis. There are complex interactions among friction, frictional heating and temperature along a fault, and the present work will be a first step towards the understanding such nonlinear interactions at high velocities.

Although still somewhat speculative, we consider that the large increase in the friction prior to the onset of visible (or massive) frictional melting (b to c in Fig. 2a) is caused by the welding of asperities due to local melting and its subsequent solidification. Enhanced junction growth of asperities in contact owing to frictional heating may be another contributing factor for the increase in friction with increasing displacement. However, we do not consider this as the major cause, since such a large peak in friction has never been observed upon step changes in the slip rate when the frictional melting is not taking place. After the peak friction is exceeded, friction continues to decrease towards the steady-state friction (c to e in Fig. 2a). This reduction in friction with increasing displacement is apparently due to increasing melt along a fault. Such general behavior of fault at high velocities has dramatic implications for the earthquake initiation processes [25]. The large increase in the frictional resistance (b to c in Fig. 2a) must act as a barrier for further acceleration of fault slip. However, once this barrier is overcome, a fault abruptly loses its frictional resistance to make a fault dramatically unstable. This loss of friction upon increasing frictional melting, together with the existence of steady-state friction at a fast slip rate, suggests that a critical slip weakening distance, D_c, may show up due to the frictional melting processes. Surface roughness of a fault and the gouge thickness have been considered to major controlling parameters of D_c [4, 5, 10, 13, 15]. Frictional melting is definitely another factor affecting D_c, a very important parameter controlling the stability of fault motion.

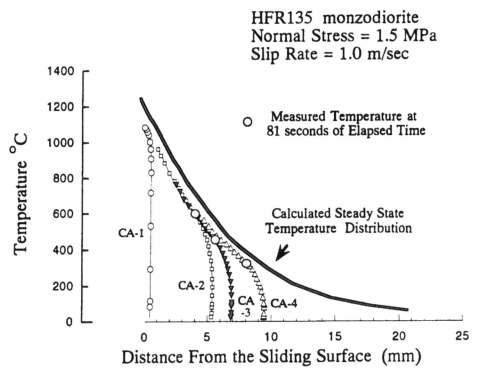

HFR135 monzodiorite
Normal Stress = 1.5 MPa
Slip Rate = 1.0 m/sec

O Measured Temperature at
   81 seconds of Elapsed Time

Calculated Steady State
Temperature Distribution

Distance From the Sliding Surface (mm)

Fig.5 Temperature as measured with four CA thermocouples (experiment of Fig. 2), plotted against the distance from the sliding surface. Solid line is the steady state temperature distribution calculated in Fig. 4(a).

REFERENCES

1. A. R. Allen. Mechanism of frictional fusion in fault zones, *Jour. Struct. Geol.* **1**, 231-243 (1979).
2. H. S. Carslaw and J. C. Jaeger. Conduction of Heat in Solids: *Oxford University Press*, 386p (1959).
3. S. P. Clark Jr. Thermal conductivity; Handbook of Physical Constants, *Geological society of America*, Memoir 97, 459-482 (1966).
4. J. H. Dieterich. Time-dependent friction and the mechanics of stick slip, *Pure Appl. Geophys.* **116**, 790-806 (1978).
5. J. H. Dieterich. Modeling of rock friction: 1. Experimental results and constitutive equations, *Jour. Geophys. Res.* **84**, 2161-2168 (1979).
6. T. L. Johnson and C.H. Scholz. Dynamic properties of stick-slip friction of rock, *Jour. Geophys. Res.* **81**, 881-888 (1976).
7. A. Lin. Glassy pseudotachylyte veins from the Fuyun fault zone, northwest China, *Jour. Struct. Geol.* **16**, 71-83 (1994).
8. D. A. Lockner and P. G. Okubo. Measurements of frictional heating in granite, *Jour. Geophys. Res.* **88**, 4313-4320 (1983).
9. R. H. Maddock. Effects of lithology, cataclasis and melting on the composition of fault-generated pseudotachylytes in Lewisian gneiss, *Scotland, Tectonophysics*, **204**, 261-278 (1992).
10. C. Marone and B. Kilgore. Scaling of the slip distance for seismic faulting with shear strain in fault zones, *Nature* **362**, 618-621 (1993).

11. D. P. McKenzie and J. P. Brune. Melting on fault planes during large earthquakes, *Geophys. Jour. Roy. Astr. Soc.* **29**, 65-78 (1972).

12. Y. Ohtomo and T. Shimamoto. Significance of thermal fracturing in the generation of fault gauge during rapid fault motion: An experimental verification, Jour. Tectonic Res. Group Japan, **39**, 135-144 (1994). (in Japanese)

13. P. Okubo and J. H. Dieterich. Effects of physical fault properties on frictional instabilities produced on simulated faults, *Jour. Geophys. Res.* **89**, 5815-5827 (1984).

14. C. T. Prewitt. Pyroxenes, Reviews in mineralogy, 7, *Mineralogical society of America*, 525 (1980).

15. C. H. Scholz. The critical slip distance for seismic faulting, *Nature*, **336**, 761-763 (1988).

16. T. Shimamoto, J. Handin and J.M. Logan. Specimen-apparatus interaction during stick-slip in a triaxial compression machine: A decoupled two-degree-of-freedom model, *Tectonophysics*, **7**, 175-205 (1980).

17. T. Shimamoto and A. Lin. Is frictional melting equilibrium melting, or non-equilibrium melting ? Jour. *Tectonic Res. Group Japan*, **39**, 79-84 (1994). (in Japanese)

18. T. Shimamoto and A. Tsutsumi. A new rotary-shear high-speed frictional testing machine: its basic design and scope of research, Jour. *Tectonic Res. Croup Japan*, **39**, 65-78 (1994). (in Japanese)

19. M. Shimoseki and H. Fujinuma. Finite element method / programming of non-steady-state Heat Flow and Stress, *Nikkan-kougyou-shinbun* (1988). (in Japanese)

20. R.H. Sibson. Generation of pseudotachylyte by ancient seismic faulting, Geophys Jour. Roy. Astr. Soc. **43**, 775-794 (1975).

21. J.G. Spray. A physical basis for the frictional melting of some rock-forming minerals, *Tectonophysics*, **204**, 205-221 (1992).

22. M. D. Teufel and J. M. Logan. Effect of displacement rate on the real area of contact and temperatures generated during frictional sliding of Tennessee sandstone, *Pure Appl. Geophys*, **116**, 840-865 (1978).

23. A. Tsutsumi, Size distribution of clasts in experimentally produced pseudotachylytes, *Jour. Struct. Geol. (in review)*.

24. A. Tsutsumi and T. Shimamoto, Frictional properties of monzodiorite and gabbro during seismogenic fault motion, *Jour. Geol. Soc. Japan*, **102**, 240-248 (1996).

25. A. Tsutsumi and T. Shimamoto, High-velocity frictional properties of gabbro, *Geophys. Res. Lett. (in review)*.

26. N. Yoshioka. A method for estimating the shear stress distribution from temperature measurements during sliding of rocks, *Jour. Phys. Earth*, **32**, 1-12 (1984).

27. N. Yoshioka., Temperature measurement during frictional sliding of rocks, *Jour. Phys. Earth*, **33**, 295-322 (1985).

Proc. 30ᵗʰ Int'l. Geol. Congr., Vol. 5 pp. 233-243
Ye Hong (Ed)
© VSP 1997

Brief Review and Preliminary Proposal for the Use of Ground Effects in the Macroseismic Intensity Assessment

ELIANA ESPOSITO[1], SABINA PORFIDO[1], GIUSEPPE MASTROLORENZO[2], ANDREI A. NIKONOV[3] AND LEONELLO SERVA[4]

1)CNR-(National Group for the Defence against Earthquakes c/o CNR-GEOMARE SUD, Via A. Vespucci 9, 80127 Napoli, Italy.
2)Osservatorio Vesuviano, Via Manzoni 249, 80123 Napoli, Italy.
3)Institute of Seismology; Joint Institute of Physics of the Earth; Russian Academy of Sciences, Blv. Gruzinskaya 10, 123810 Moscow, Russia.
4)ANPA, National Agency for the Protection of the Environment. Via V. Brancati 48, Roma, Italy.

Abstract

Ground effects have been an integral part of intensity (I) assessment since introduction of the first scale by de Rossi and Forel in 1883. However during the first half of this century an enormous amount of work has been devoted to improving I scales through a detailed classification of the different types of buildings and expected damage. As an extreme culmination of this tendency, the recent up-dated EMS 92 macroseismic scale completely ignores ground effects, only listing them in a short Appendix. Although this approach is reasonable for densely urbanized areas, it does not reflect the full meaning of the I parameter, it does not work for I assessment in the highest degrees of the scales, it is unsuitable in remote areas and cannot be used for comparison of historical events and seismic hazard evaluation in areas affected by strong events with return periods longer than the historical records. On the other hand, worldwide studies on the empirical relations between seismic parameters and ground effects demonstrate that, even if type and relevance of surface effects strongly depend on the local geomorphological setting, their definition allows a realistic estimation of the earthquake size. In this paper these relations are briefly discussed and preliminary proposals are given for the better use of ground effects in I assessment.

Keywords: Earthquake, Intensity, Ground effects, Macroseismic scales.

INTRODUCTION

Intensity (MCS: Mercalli-Cancani-Sieberg, MSK: Medvedev-Sponheuer-Karnid, MM: Modified Mercalli, Japanese, scales; the first three are here considered equivalent) is still widely used in the world as a measure of the earthquake size and for engineering purposes. As an example the Chinese Building Code [4] contemplates the evaluation of the susceptibility to liquefaction of the foundation material through many parameters; among which the Standard Penetration Test (SPT) is derived as a function of earthquake I.

Historically the most appropriate set of parameters to be included in I evaluation has been object of continue intense debate. However, since the beginning of the research for a standardized I scale [7, 18] the effects considered included: a) the human behavior, b) damage to buildings, c) effects on the natural environment (the so called ground effects).
On the basis of this methodological approach, several scales were developed with differences

related mostly to the spectrum of structure types, to be usefully adaptable to different countries and cultural environments. Nevertheless, the ground effects, that are the only ones not changing with the cultural environment, remain subordinated in I assessment and in most earthquake reports they are poorly considered or even ignored. At the top of this tendency, the last Macroseismic Scale proposed by the European Seismological Commission [12] considers only the human behaviour and the damage to an extremely detailed buildings type. Ground effects are discussed in a short Appendix, but not included expressively in the scale.

However it is important to point out that the assessment of the I value without taking into consideration the ground effects is not correct because it does not allow: a) a proper estimation for the highest degrees; b) an I evaluation in scarcely populated areas; c) a comparison with historical events.

Furthermore, the seismotectonic analyses carried out for seismic hazard estimation needs a long record (i.e. at least the entire Holocene) of seismicity. Ground effects can be recorded in the geomorphological setting and therefore recognized by the specialists (e.g. paleoseismological studies, in the sense of Michetti [19] and Serva [24]).

This paper, that should be regarded as an initial stage of the research, first briefly summarizes some historical and modern evidence of important ground effects induced by earthquakes, and than traces a possible framework for the proper use of ground effects in the I assessment.

PRIMARY AND SECONDARY GROUND EFFECTS INDUCED BY MAJOR EARTHQUAKES

The primary effects, directly caused by earthquakes, include significant ground deformations such as surface faulting, uplift and subsidence. An ample literature describes extensive surface evidence of landscape modifications directly related to seismic events [13, 25, 26, 28, 33]. In general, primary effects, at various ranging scales, are always associated with major earthquakes (magnitude greater than 6) [e.g. 2]. The San Andreas active (capable) fault is the most remarkable example of a site where primary ground features related to great earthquakes occur. Other remarkable examples are reported for the 1964 Alaskan earthquake [14]; the 1920 Haiyuan and the 1970 Tonghay-Yunnan (China) earthquakes [32, 35, 36]; the 1891 Mino-Owari (Japan) earthquake [20], where in the village of Neodani a museum has been recently constructed to expose the magnificent example of surface faulting; the 1976 great Guatemala event [21]; the 1992 Lander (California) earthquake [15] and the 1915 Fucino (Italy) earthquake [22].

The short-term secondary ground effects of earthquakes include liquefactions, landslides and fractures.

Classical examples of seismically induced liquefaction are the 1811 New Madrid (Missouri) and the 1886 Charleston (South Carolina) events. The latter caused over one hundred liquefaction phenomena within a 1500 squared kilometers area [7]. Also one of the strongest Italian earthquakes, occurred on February 1783 in the Calabria Region (Southern Italy), triggered several different geomorphological phenomena such as: fractures, landslides damming the rivers, and extensive liquefactions. In particular, the river diversions or damming produced at least 215 lakes ranging in size between some hundreds to 1000 squared meters; furthermore, the istantaneous drying-up of some rivers, followed by flood waves and eruptions of ground water with characteristic sand volcanoes due to liquefaction, were observed[5].

As clearly discussed in Keefer [17] earthquakes are recognized a major cause of landslides. The "Chronicle on Bamboo Slip" is the first known document on an earthquake-induced landslide dated back to the 1767 B.C. [34]. The 760-750 B.C. The Jerusalem earthquake, mentioned in the Bible, represents the most ancient known event inducing landslides in the Mediterranean area [13].

Fissures in the ground are also a common feature in the earthquake reports, e.g.[3,6].

INTENSITY EVALUATION FROM PRIMARY AND SECONDARY GROUND EFFECTS

Numerous relationships between primary effects (surface rupture length, maximum and average displacement) and magnitude have been derived in the last decades. The last developments are reported by Wells and Coppersmith [31].

The other primary effects (uplift or subsidence) are, in a way, considered in the relationships slip-rates-magnitude (i.e.[1,27]). Magnitude vs. rupture area and rupture width [31] can be also considered in this context.

Until now however, no such relationships have been developed with I, although an abrupt manifestation of ground effects is normally recognized at higher values of intensity ($I \geq X$, see Tab. 1, [22]). Furthermore, if I is an estimation of the earthquake size for events having similar hypocentral depths it should increase with magnitude. In other words, an earthquake of M=6.5-7.0 should produce a lower I of an earthquake of M=7.5-8.0. For instance, Johnstone [16] corroborates earthquake size estimates for the great intraplate earthquakes of New Madrid (USA) 1811-1812, Charleston (USA) 1886, and Lisbon 1755 with analysis of

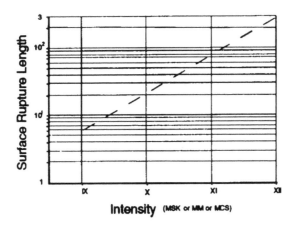

Fig. 1 First attempt of regression of surface rupture length on intensity (crustal earthquakes).

landslides, liquefaction and tsunami occurrence. Thereafter, it would be useful to construct relationships between I and primary ground effects.

Table 1   Summary table of the ground effects described in the intensity scales. (from Serva [23], modified).

	MSK	MM	MCS	JAP
Cracks in saturated soil and/or loose alluvium				
up to 1 cm	VI			
a few cm	VIII	VIII	VIII	
up to 10 cm	IX	IX		
a few dm up to one m	X		X	
Cracks on road backfills and on natural terrigenous slopes over, 10 cm	VII, VIII IX	VIII	VIII	
Cracks on the dry ground or on asphalted roads	VII, IX XI		X, XI	VI
Faults in terrigenous terrains	XI		XI	
Fault in rocky terrains	XII			VI
Liquefation and/or mud volcanoes and/or subsidence	IX, X	IX, X	X, XI	
Landslides in sand or gravel dykes	VII, VIII, X	VII	VII	
Landslides in terrigenous slopes	VI, IX X, XI	X	X, XI	VI, VII
Rockfalls	IX, XI, XII	XII	X, XI	
Clouding in the closed water bodies and formation of waves	VII, VIII, IX	VII	VII, VIII	
Formation of new water bodies	VIII, X, XII		XII	
Change in the direction of flow in water courses	XII		XII	
Flooding	X, XII	X	X	
Variation in the water level of well and/or the flow rate of springs	V to X	VIII	VII, X	
Springs which are drying out or are starting by gush out	VII, VIII, IX			

The one reported here (Fig. 1) is derived from the Wells and Coppersmith [31] relationship among surface rupture length and magnitude. Intensity values have been obtained through an analysis of the worldwide I - Magnitude relationships associated with our experience. Fig. 1 should be read only for deriving order of magnitude of the surface rupture length. The I scale saturates at XII degree, therefore rupture lengths greater than 200 Km should be classified with the same I.

We are conscious that fig. 1 should represents just a starting point. Much work is needed to obtain a representative dataset for a reliable relationship. The scope of its publication is therefore to make a gentle provocation for the Scientific Community to work in this line.

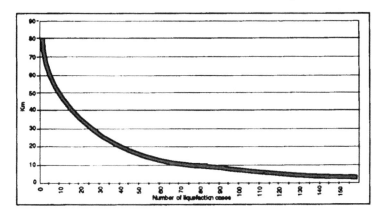

Fig. 2   Graph showing the number of liquefaction cases versus epicentral distance. All of the historical liquefaction cases have been plotted as a function of their distance from the epicenter. The resulting curve shows that only 10% of cases occurred at a distance greater than 40 km. (from Galli P. & Ferreli L. [11], redrawn).

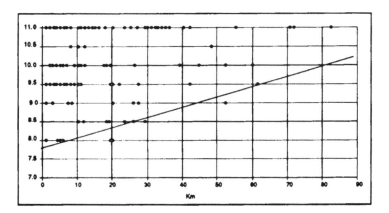

Fig. 3   Graph showing epicentral distance versus epicentral intensity. The epicentral distance of liquefaction events is directly linked to the earthquakes intensity. The area below the line that has no points could represent those sites with a very low probability of experiencing liquefaction relative to determined seismic source areas. (from Galli P. & Ferreli L. [11], redrawn).

Regarding the secondary effects, according to Tinsley et al. [29] significant liquefaction phenomena have been observed at magnitudes as low as 5.0-5.5, but are more characteristic of higher magnitudes. Furthermore, these authors have shown that for distances exceeding 100 km, magnitude 7.5 seems to be a minimum threshold for liquefaction (see also [30]). Galli and Ferreli [11], on the basis of all published data relative to the Italian earthquakes from 453 B.C. to 1982 A.D., integrated with a theoretical attenuation model, recognized some relations between earthquake I and occurrence of liquefaction. In particular, the main result is that liquefaction is induced for a minimum site intensity of VII MCS. The maximum epicentral distance at which liquefaction occur is directly linked to the earthquake I and only 10% of cases occur at a distance greater than 40 km (Fig. 2, 3).

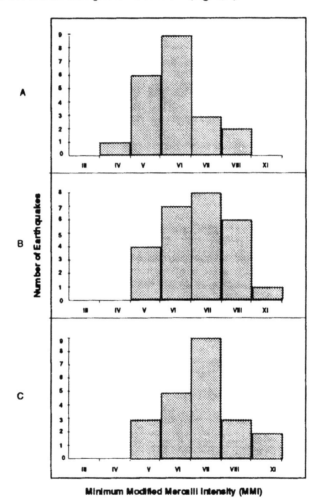

Fig. 4   Minimum Modified Mercalli Intensity at which landslides occur. A. Minimum intensities for disrupted slides or falls. B. Minimum intensities for coherent slides. C. Minimum intensities for lateral spreads or flows (from Keefer [17], redrawn).

Keefer [17] collected 40 worldwide major historical earthquakes ranging in magnitude between 2 and 9.2. This work shows two important results. The first is that the area interested by landslides could vary from 0 squared kilometers at M= 4, to approximately 500,000 squared kilometers at M= 9.2. The second is reported in Fig. 4; for details see the figure captions. In addition, according to the same author only few earthquakes have reactivated older landslides, being generally newly created. Different results have been obtained by Esposito et al. [9,10]. In particular, from the study of many landslides induced by the 1980 Irpinia (Italy) earthquake (M= 6.5), most occurrence, over an area of 20,000 squared kilometer, were reactivations of dormant landslides. In the same paper (Fig. 5), four distinctive types of slope movements were identified (rock falls, 47%; slump-earth flows, 40%; other types 4%; not defined origin 9%). A major result is that the number of earthquake induced landslides (in this case, because of the geomorphological setting, especially the rock falls) for squared kilometer, may be a good indicator of I distribution.

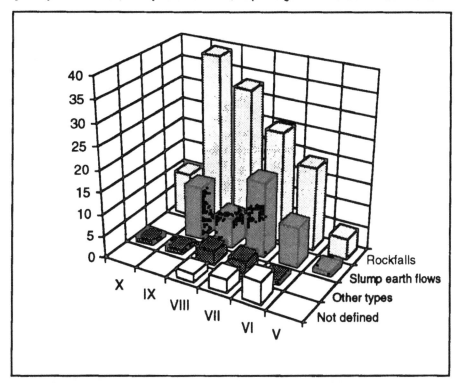

Fig. 5 Irpinia (southern Italy) earthquake. Distribution of 200 landslides (of different type) in the various intensity levels. Rockfalls, because of the geomorphological setting, are the most informative features. Most of the landslides fall within the area countered by the VII isoseism; in this area paleoseismic research, based on landslides, can produce significant results.

Fractures in the ground are another important secondary effects. Fig. 6, taken from, Carmignani et al. [3], shows, for the Irpinia earthquake, a good correspondence between areal distribution of ground fissures density and the VIII and IX isoseismal lines.

Nikonov carried out a systematic study of the ground phenomena observed in about 100 earthquakes occurred between 19th and 20th century in three major regions of the ex USSR territory , namely the Northern Central Asia, the Caucasus and the Crimea. These effects were classified into four groups with increasing relevance (for details see Tab. 2). This table also shows the minimum I value at which the different types of ground effects are generated. The main results are: 1) relevance of ground effects (number and type) are positively correlated with the MSK-64 I scale; 2) light ground effects begin at I: IV-V; 3) considerable effects in the terrigenous soil begin at IV-VI; 4) Most of detected relief changes and rock disturbances were grouped within VII and VII-VII isoseismal lines.

1980 IRPINIA (or Campano - Lucano) EARTHQUAKE

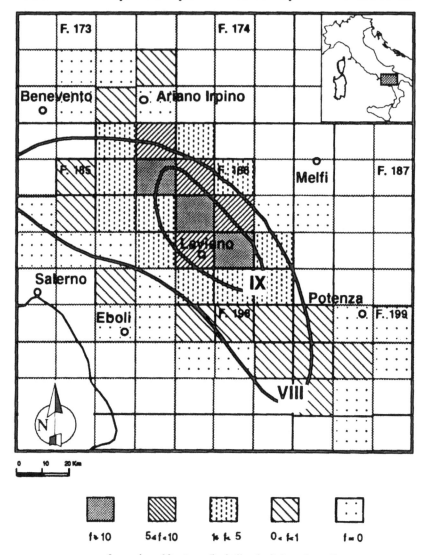

f = number of fractures (including faults) per linear K m

Fig. 6 Areal distribution of the frequency of fractures in the studied area. The mesh corresponds to the limits of the IGM official topographic maps. The thicker lines show the isoseismal of the VIII and IX grade (from Carmignani et al., modified [3]).

More in general, these data show that ground effects should appear in the MSK scale 1-2 degrees before the degrees they start to be considered. For example, liquefaction should be included in VII degree while now it appears in the IX.

Table 2 Relation among seismogravitational deformations and intensity. CA: mountain regions of northern Central Asia; C: Caucasus, Cr: Crimea.

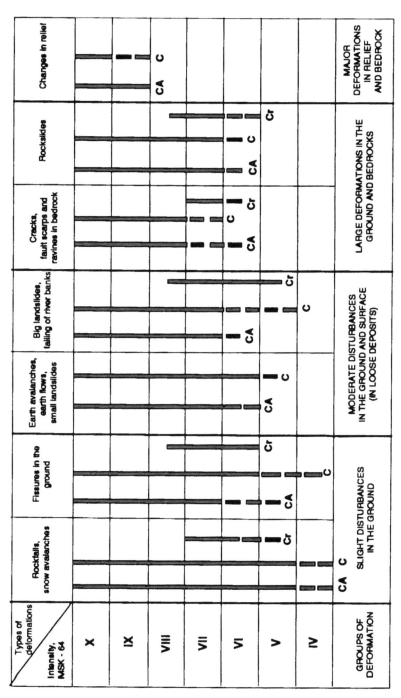

Recently, some attempts have been made to classify the ground effects in a systematic way for I assessment. In particular Dengler and MacPherson [6] have utilized many of these effects as damage indicators in assessing I for scarcely populated areas.

We think, however, that these works should continue in order to categorize these effects in the scale in a way comparable to what has been done for the effects on the buildings. The schema traced here hopefully represents a useful starting point in this line.

CONCLUSIONS

The present review shows that although a worldwide experience confirms the relevance of the ground effects in defining the earthquake size, a very small consideration has been given to them until now. In the perspective of a global approach to seismicity, which addresses a reconsideration of the I parameter and the understanding of the deterministic relations between earthquakes and ground phenomena, a standard criterion for assessing I should be defined. Such an aim implies the improvement of the quantitative descriptions of the ground phenomena following unique and adequately detailed schemes (for example the classification proposed by Nikonov or other equivalent) and carrying out a routinary comparison between the macroseismic degree and a standard set of phenomena. In this sense, the proposals traced in this paper may represent a use ful starting point. A possible reference for the collection and coordination of the different contributions could be an International Commission for standardization of macroseismic scales.

Acknowledgments

Thanks are due to Eutizio VITTORI and Alessandro Maria MICHETTI for their helpful comments and suggestions.

REFERENCES

1. J.G. Anderson, S.G. Wesnousky, M.W. Stirling, Earthquake size as a function of fault slip rate. *Bull. Seim. Soc. Am.* **86**, 683-690 (1996).
2. B.A. Bolt. *Earthquakes.* W.H. Freeman, San Francisco, (1978).
3. L.Carmignani, G. Cello, A. Ferroni Cerrina, R. Funiciello, O. Kalik, M. Meccheri, E. Patacca, P. Pertusati, G. Plesi, F. Salvini, P. Scandone,I. Tortorici, G. Turco. Analisi del campo di fratturazione superficiale del terremoto campano-lucano del 23.11.1980. Rend. Soc. Geol. Ital., **4**, 451-465 (1981).
4. Chinese Building Code. Earthquake resistant design code for industrial and civil buildings. TJ11-74. China Build. Pubbl. House, (1974).
5. V. Cotecchia, A. Guerricchio, G. Melidoro. The geomorphogenetic crisis triggered by the 1783 earthquake in Calabria (Southern Italy). Bari, Italy. *IAEG-AIGI Conf. Proceed.*, **6**, 245-304 (1986).
6. L. Dengler & R. McPherson. The 17 August 1991 Honeydaw Earthquake North Coast California: a Case for Revising the Modified Mercalli scale in Sparsely Populated Areas. *Bull. Seis. Soc. Am.* **83**, 4. 1081-1094 (1993).
7. M.S. de Rossi. Programma dell'osservatorio ed archivio centrale geodinamico. *Boll. Vulc. Ital.* **10**, 3-124 (de Rossi-Forel scale, pp. 67-68) (1883).
8. Ebasco Services Inc. Paleoliquefaction features along the Atlantic Seabord. USNRC NUREG/CR-5613 (1990).
9. E. Esposito, A. Gargiulo, G. Iaccarino, S. Porfido. Analisi dei fenomeni franosi in aree ad elevata sismicità in Appennino meridionale. Proceed., *Accademia Nazionale Lincei, Roma, Italy " La stabilità del suolo in Italia: zonazione della sismicità- frane, Conf.PRoceed..* (in press)
10. E. Espotito, A. Gargiulo, G. Iaccarino, S. Porfido. Distribuzione dei fenomeni franosi riattivati dai terremoti dell'Appennino meridionale. Censimento delle frane del terremoto del 1980. *CNR-IRPI, Alba (CN), Italy. "Prevention of hydrogeological hazards: the role of scientific research" Conf. Proceed.* (in process)
11. P. Galli & L. Ferreli. A methodological approach for historical liquefaction research. In: *AEG Special*

Publication "Perspectives in Paleoseismology" L. Serva and D.B. Slemmons Eds., 6, 35-48 (1995).

12. G. Grunthal (Ed.) European Macroseismic Scale 1992 (up-dated MSK Scale). *CONSEIL DE L'EUROPE, Cahiers du Centre Europeen de Godynamique et de Seismologie,* 7, Luxembourg, 79pp (1993)

13. E. Guidoboni, A. Comastri, G.Traina. Catalogue of ancient earthquakes in the Mediterranean area up to the 10th century. *ING-SGA,* Bologna, Italy (1994)

14. W.R. Hansen. 1966, Effects of the earthquake of March 27, 1964, at Anchorage, Alaska. *U.S. Geol. Serv. Professional Paper* 542-A (1966)

15. E.W. Hart, W.A. Bryant, J.A. Treiman. Surface faulting associated with the June 1992 Landers earthquake, California. *California Geology,* 10-16 (1993)

16. A.C. Johnston. Seismic moment assessment of earthquake in stable continental regions-III. New Madrid 1811-1812, Charleston 1886 and Lisbon 1755. *Geophys. J. Int.* 126, 314-344 (1996)

17. D.K. Keefer. Landslides caused by earthquakes. *Geol. Soc. Am. Bull.* 95, 406-421 (1984)

18. G.Mercalli. Sulle modificazioni proposte alla scala sismica de Rossi-Forel. *Boll. Soc. Sism. Ital.* VIII, (1902)

19. A.M. Michetti. Coseismic surface displacement vs. magnitude: relationships from paleosismological analyses in the Central Apennines (Italy). *Journal Geodetic Soc. Japan, CRCM '93, Proceed.,* 375-380 (1994)

20. J. Milne & W.K. Burton. *The great earthquake in Japan. Lane, Crowford e Co.* Jokohama, Japan (1892)

21. G. Plafker, M.G. Bonilla, S.B. Bonis. The Guatemala earthquake of February 4, 1976, a Preliminary Report, A.F. Espinosa (Ed.) *U.S. Geol. Surv. Professional Paper,* 1002, 38-51 (1976)

22. L. Serva, A.M. Blumetti, A.M. Michetti. Gli effetti sul terreno del terremoto del Fucino (13.1.1915). Tentativo di interpretazione della evoluzione tettonica recente di alcune strutture. *Mem. Soc. Geol. It.* 35, 893-907 (1986)

23. L. Serva. Ground effects in intensity scales. *Terra Nova,* 6, 414-416 (1994)

24. L. Serva. Criteri geologici per la valutazione della sismicità: considerazioni e proposte. Accademia Nazionale Lincei, "Terremoti in Italia: *Previsione e prevenzione dei danni"* *Conf. Proceed.* 122, Roma, 103-116 (1995)

25. K.E. Sieh. A review of geological evidence for recurrence times of large earthquakes. *AGU Earthquake Predict. - An Intern. Rev.,* Maurice Ewing, 4, 181-207 (1981)

26. D.B. Slemmons. Faults and earthquake magnitude. U. S. Army Corps of Engineers, Miscellaneous papers S-73-1, 6 (1977)

27. D.B. Slemmons & C. de Polo. Evaluation of active faulting and associated hazards. *"Active tectonics",* National Academy Press Washington D. C., 45-62 (1986)

28. D.P. Schwartz & K.J. Coppersmith. Fault behaviour and characteristic earthquakes: examples from the Wasatch and San Andreas fault zones. *Journal of Geophys. Res.,* 89, 5681-5698 (1984)

29. G.C. Tinsley, T.L. Yud, D.M. Perkins, A.T.F. chen. Evaluating liquefaction potential. In J.I. Ziony (Ed) "Evaluating earthquakes hazards in the Los Angeles retion-an Earth Science Perspective. *U.S. Geol. Surv. Professional Paper* 1360, 263-316 (1985)

30. E. Vittori, S. Sylos Labini, L. Serva. Paleoseismology: review of the state-fo-the-art. *Tectonophysics,* 193, 9-32 (1991)

31. D.L. Wells & K.J. Coppersmith. New empirical relationships among magnitude, rupture length , rupture width, rupture area, and surface displacement. *Bull. Seis. Soc. Am.,* 84, 4, 974-1002 (1994)

32. Yang Bin, Zhou Junxi, Zhang Jie, Wang Yanbin, Yuan daoyang, Shi Xiaofei, Xhen Yukun, Wang Fei, Hou Kangming. Seismogeology and quaternary geology along northern margin of the Qinghai-Xizang plateau. *30th IGC, Beijing (China) Field Trip Guide T 356,* Geological Publishing House (1996).

33. T.L. Youd & S.N. Hoose. Historical ground failures in Northern California triggered by earthquakes, *USGS Professional Paper,* 993 (1978).

34. F. Xue-Cai & G. An-Ning. The principal characteristic of earthquake landslides in China. Bari, Italy, *IAEG-AIGI Conf. Proceed.,* 2, 27-45 (1986).

35. G. Zengjian & Q. Baoyan. Prediction of dynamic hazard in seismic areas. Bari, Italy, *IAEG-AIGI Conf. Proceed.* 1, 1p. (1986).

36. W. Zhong-Qi. 1986, Evaluation on seismic effect of faults and ruptures. Bari, Italy, *IAEG-AIGI Conf. Proceed.,* 5, 173-212 (1986).

Proc. 30ᵗʰ Int'l. Geol. Congr., Vol. 5 pp. 245-253
Ye Hong (Ed)
© VSP 1997

A Nonlinear Model for Intraplate Earthquake

LI LI[1] , ZHANG GUOMIN[1] AND SHI YAOLIN[2]
1) Center for Analysis and Prediction, SSB, Beijing 100036, China
2) Graduated School, USTC, Beijing 100039, China

Abstract

Intraplate earthquakes are very hazardous disaster for human's society. The features of earthquake-generating process are discussed in this paper by using a dynamic nonlinear model with Chinese Continental area as an example. The theoretical computation result of the model shows similarity to the real seismicity in Mainland China. The earthquakes in the model occur periodically. The shocks in the system break out in seismic belts. In different active period the main active belts are different. Although the variation and stress images of the system are very complicated, there are some omen observed in the stress filed for the next abruption of the system to happen.

Keywords: intraplate earthquake, geodynamics, nonlinear model.

INTRODUCTION

Earthquakes are usually divided into two groups: intraplate earthquakes, and interplate earthquakes. Most of the intraplate earthquakes are located in the continental area. Those earthquakes could do great harm to human's society because they often take place in most populated regions. The number of the continental shocks is only 15% of the whole global events, but 85% of the global hazards are from continental shocks. About 1/3 continental earthquakes occured in China. In this century the casualties of earthquakes in China make up 1/2 of those in the whole world. The study on intraplate earthquakes is very important for hazard reduction in China.

The dynamics of earthquake-generating process has been discussed for many years. The model proposed by Burridge and Knopoff [1] with a spring and a slider as elements was the early example of its kind. Afterwards, Byerlee[2], Dieterich[4], Rice et al.[9] calculated a one-degree system with only one slider. In 1991, Zhu Yuanqing and Shi Yaolin [11] modeled the seismic activity in several parallel fault zones using a system with a few non-linear coupled elements. Recently, Shi Yaolin, Zhang Guomin and Geng Luming[10] have done more profound research in this field.

By using a dynamic nonlinear model, the genetic features of the preparation and development of earthquakes are discussed in this paper based on the seismic data in Chinese continental area.

DESIGN OF THE DYNAMIC MODEL

Mainland China is located in the southeastern part of Eurasia Plate. The crust under Mainland China is composed of tectonic blocks with different classes. Based on crustal structures, seismotectonics and regional seismic activities, Gao Weiming et al[5]. divided Mainland China into 31 seismic belts and more than 70 seismogenic zones. Based on the above mentioned geological nature, the seismic regions in China are discussed as a big seismogenic system which consists of several seismic belts. In each belt there are several strong seismogenic bodies, shown as Fig.1a. Before the mechanical model on Fig.1a are computed, the following points should be assumed:

(1) During the whole seismogenic process, the accumulation of the stress and the strain reflects the accumulation of rheological bodies, because the preparation of a strong shock may need several hundreds to thousands of years or even longer. On the other hand, the occurrence of an earthquake is a rapid process of unstable rupture within a very short time. During this short period seismic sources breaks, rock bodies slide and stress drops, and the wall rock is affected. So it can be seen as a discontinuous and nonlinear process. In order to exhibit all these features, the sources are simplified as a Maxwell's body cascaded with a rigid slider, which is named " basic organ ".

(2) The whole geological region is treated as a united seismogenic system with several seismic belts, in each belt there are several seismogenic bodies. In the model, seismic belts are simplified as several paralleled basic organs, and several belts make up the whole seismogenic system.

(3) A coupling organ, which consists of a spring and a damping, is placed between two basic organs. It represents the coupling within a seismic belt and between two belts.

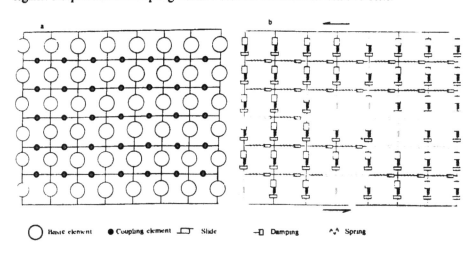

○ Basic element   ● Coupling element ⊏⊐ Slide   ⊸ Damping   ⌁ Spring

(a) the 6×8 model   (b) the abstract model
Fig.1 Abstract block diagram of intraplate seismotectonics

The mechanical model is shown in Fig.1b. Considering the constraint of the condition and the ability of calculation equipment, we chose the 6×8 model for a big seismogenic system, in which there are 6 seismic belts, and 8 seismogenic bodies in each belt.

Mainland China consists of a series of sub-class blocks. Under the force from the India Plate and the Pacific Plate, this continental area act as a large seismogic system, consisting of many seismic belts and many seismic sources. Although the model shown on Fig. 1 is a very simplified one, it presents the basic features of the Chinese intraplate earthquakes.

DYNAMIC SYSTEM OF EQUATIONS AND ITS SOLUTION

The theoretical solution of the mechanical system shown in Fig.1 has been given by Shi Yaolin.The mechanical features of the long term seismogenic stage of a source and the characteristics of the rapid breaking stage are different, so the mechanical modeling and calculation for a seismic source should be done at seismogenic stage (the accumulating of stress and strain) and at seismic stage (the slider moving within a short time), respectively.

The Stress Accumulation Stage(Seismogenic Stage)
The model is assumed to have m×n basic organs and (m-1)×(n-1) coupling organs. During this stage, each basic and coupling organ follows the constitutive equations for a Maxwell's body:

We assume that:
(1)The compressive stress is positive, the compressive stress is negative,the clockwise shear force is positive and the anticlockwise shear force is negative.
(2)The total strain rate of cascaded organs is up to the sum of the strain rate on each organ.
(3)The connecting points in the model keep balanced under a constant strain rate, i.e., the sum of stress on all the organs with a same connecting point is zero.
Based on the above assumptions,

When the initial stress values are given, the stress value of each organ can be calculated[7].

Slider Moving Stage(Seismic Stage)
A Tectonic shock is a kind of mechanical destabilization which can be treated as a limited sudden dynamic sliding on the section caused by arbitrary small asymptotic changes of regional stress or displacement. Burridge (1973) had given the strength limits of finite stress and the destabilizing criterion.

In our modeling, the destabilizing criterion of each slider is simplified [7]. It is assumed that the total strain alignment of the cascaded organs is equal to the sum of the strain alignment on each organ, and the stress increment of each organ follows the balance of connecting spots, i.e., the sum of the stress increment of the organs with a same spot is zero, and the total strain of the whole system would not change at the moving moment, that is:

$$\Delta \varepsilon = \sum_{j=1}^{n} \Delta \varepsilon_{ij}$$

Using the above method[7], the solution of a set of linear equations can give the stress alignment of each organ. In the whole modeling process, when the initial stress values are given, the stress of each organ at each time point can be calculated. When the stress of the organ meets the sliding stress level, the linear set of equations can give the stress value after the stress adjustment. Then, the new stress value is taken as the initial stress to be calculated..

THEORETICAL RESULTS AND REAL SEISMIC ACTIVITIES

The theoretical results of the model can be obtained with the above mentioned calculation. Considering that the difference of medium features among the crustal rocks is small, the elasticity modules and the viscosity coefficient used in this modeling for each organ are the same. The static frictional strength of each organ is different to each other, which shows the difference of broken   strength in the organs. The ratio of dynamic frictional strength and static frictional strength is 1:1. 25. The boundary condition has a constant strain rate.

Seismic Features of Mainland China

Fig.2 is the cycle image of Mainland China during this century given by Qiu Jingnan[9], in which there are 4 cycles for the strong earthquakes. It shows that each cycle contains an active stage and a quite stage.

Zhang Guomin et al.[11] pointed out that the frequency of strong earthquakes during each active period in Mainland China has an index change with time, shown by Fig.3. The analysis on the spatial distribution of shocks occurred in China during this century shows that the main active regions in different seismic cycle are different. To counter the above temporal and spatial features of earthquakes in Mainland China during this century, the model mentioned above is computed.

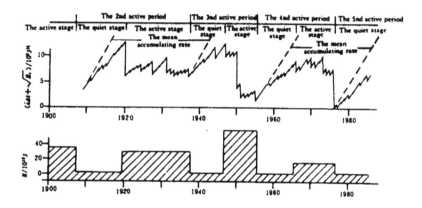

Fig.2 Schemmatic map showing 4 cycles in China in this century (Qiu et al., 1994)

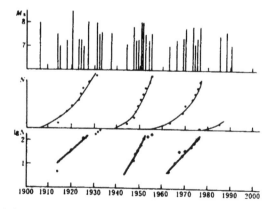

Fig.3 Index analysis for strong earthquake activities in Mainland China.

Seismic Features of the Model

Fig.4 displays the calculated curves of the model. The time interval is chosen from t=980 to t=1160, where the time t is a relative parameter and it do have unit. During the time interval, the shocks occurred in the system can also be divided into 4 stages, i.e., 4 cycles, such as: I $t\in[980, 1040]$; II $t\in[1040, 1090]$; III $t\in[1090, 1120]$; IV $t\in[1120, 1160]$. Comparing the curves in Fig.5 to those in Fig.2 and Fig.3, it can be found that the theoretical results from the model are very similar to the real seismic features.

Fig.5 shows the distribution of shocks during each cycle in the model. The 6×8 blocks represent 48 basic organs, the number on the upper horizontal line and near the left vertical line give the latitude and the longitude of the organs. The left figure shows the sequence of shocks and the right one shows the magnitude and the distribution of the shocks. It can be seen that the main active belts in one cycle are different with those in other cycles, and with the time passing, the main active belts transmit significantly. The transmission follows some rule, during cycle I, belt 1→belt 6; during cycle II, belt 5→belt 1; during cycle III, belt 3→ belt 2→belt 5; during cycle IV, belt 6→belt 2. Since the boundary condition of the model is designed as a confined constant strain rate, the transmission of main active belts can help us to understand the similar transmission in Mainland China.

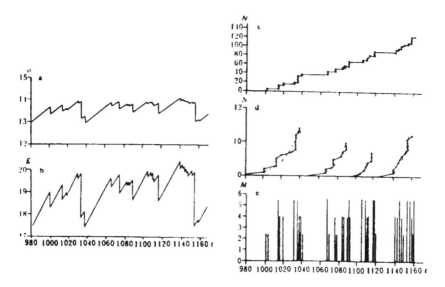

Fig. 4   When $t\in[980, 1160]$ (a) mean stress curve; (b) mean strain energy curve; (c) released energy curve; (d) accumulated frequency curve; (e) M-t map

The Stress Image of the model

Fig.6 is a picture that shows the stress changes with time during the time interval from 980 to 1040 in our 6×8 model. The numbers in the figure are relative stress values which show the stress level of each seismogenic area. The relative stress can be obtained

from: $V_{ij} = \dfrac{\sigma_{ij} - \tau_{ij}^{d}}{\tau_{ij}^{s} - \tau_{ij}^{d}} \times 10$,   where σ_{ij} is the stress on the organ, τ_{ij}^{d} and τ_{ij}^{s} is its dynamic

frictional strength and static frictional strength respectively. When and only when $Vi,j=10$,

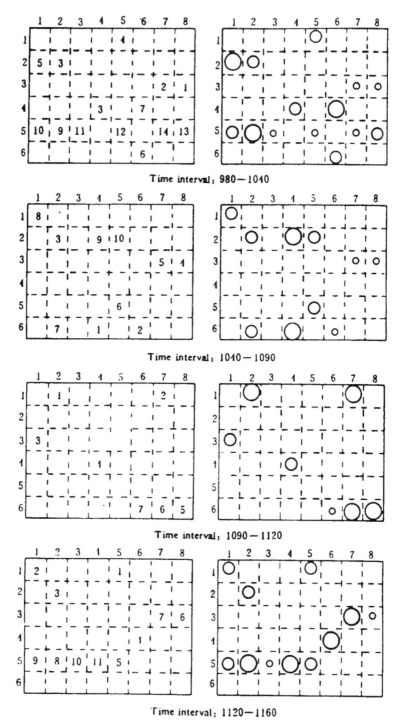

Fig. 5 The spatial distribution of ruptures in each cycle given by the model

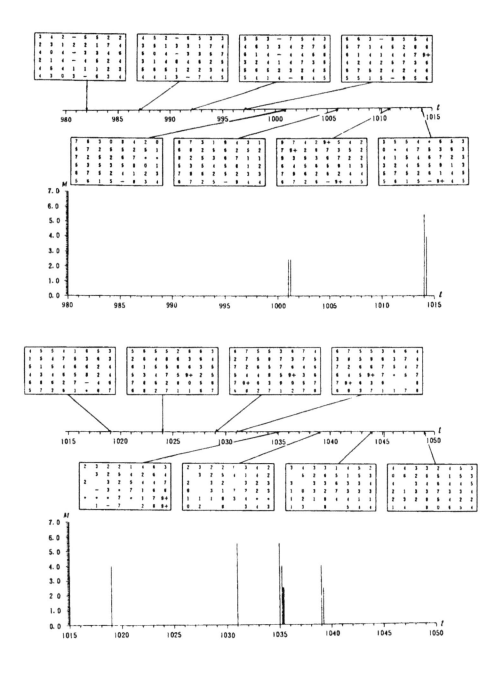

Fig.6 The stress image of the system within [980, 1150]

the organ breaks. In Fig.6 , "–" means the stress in this organ is lower than its dynamic frictional strength, "9+ " means the relative stress on this organ is larger than 9 and smaller than 10 and the organ isn't broken, and "•" means the organ breaks. The figure shows that when t=982, the stress level of the whole system is low, but a series of abruptions construct the nonlinear stress changes of the whole system after a period of time, which is due to the heterogeneity of the organs and the whole system. In detail, after 15 time points, the stress level of the whole   system increases under the constant boundary strain rate. When t= 997, the relative stress on organ $E_{3,8}$ was increased and $E_{3,7}$ was broken. The above two abruptions cut down the stress on $E_{6,6}$. During the time interval from 1001 to 1011, the stress adjustment of the whole system and the effect of boundary strain rate make the stress accumulated on each organ, and at t=1011, the relative stress values of $E_{1,5}$, $E_{2,2}$, and $E_{6,6}$ are higher than 9. When t=1019, $E_{6,6}$ broke. Since the system is under a constant boundary strain rate, the accumulation of stress, strain and strain energy is continuous. Therefore, before t=1035, the stress and strain energy of the system   had been accumulated to a high level, then when t=1035, 5 abruptions took place one by   one, which means the seismic climax of this interval. And just because of the 5 abruptions, the stress, strain and strain energy of the system released rapidly. So, after t=1035, the stress of the system dropped to a low level, which leads to a long-time stress seismogenic stage after t=1040. Then a new seismically active stage began. It can be observed that after an organ is broken, the stress level of the whole system drops. But the stress on the organs which are at the same belt of the broken organ is increased, and the stress increments decrease with the increase of distance from the broken organ.

The organs which are not on the same belt with the broken one have their stress cut down, the stress drops decrease with the increase of the distance from the broken organ. Special attention should be paid to that when the relative stress on an organ in larger than 9 and near 10, it may be broken, but it doesn't have to be broken. For instance, when t=1035, the relative stress values on $E_{5,6}$ and $E_{6,8}$ reach "9+", $E_{5,8}$ was broken at first, and $E_{6,8}$ was caused to be broken. So the stress on $E_{6,8}$ which is not at the same belt with the broken organ, decreased rapidly and strongly. And $E_{6,8}$ was not broken during this whole time interval. It can also been found that the stress patterns of the system are very complicated.

DISCUSSION

During the modeling and the calculation, we notice that in each cycle the total stress of each belt has a same value at the same time point. In some belt, the stress distributes approximately evenly on each organs. There are few shocks in these belts. These belts are called quite belts. At the same time, in some belts although the stress level are the same as the quiet belts, the stress in them   are concentrated to several organs. In these belts there are strong and continuous shocks. These belts are called active belts. It is suggested by this model that whether the belts are active or not in a cycle mainly depends on the stress distribution of these belts.

Fig.6 shows that the stress image at the beginning of a cycle are possibly to be used to predict the broken spots within this cycle. For instance, the 3 organs whose relative stress is higher than 9 when t= 1011 all were all broken in this cycle. This   figure also shows that the stress image at a definite time can predict the next broken spot. For example, when t=997, the relative stress on $E_{3,8}$ is higher than 9, and when t= 1001, this organ is broken.

Despite the above-mentioned success, it should be pointed out that this dynamic nonlinear model is only in its primary test stage. The results are far from a satisfactory precise prediction. For example, when the relative stress value of an organ is higher than 9, it doesn't have to be broken and it even won't break during the whole cycle. The more sophisticated study is required in future work.

Aknowledgements

We would like to thank Geng Lumiu and Liu Jie for their valuable help in this study.

REFERENCE

1. R Burridge, et al., Model and theoretical seismicity. B.S.S.A., 57:341-371, (1967).
2. D.J.Byerlee. Friction of rocks. P.A.G., 116:615-626, (1978).
3. Chen Huikai, Linear network and system. Translated by Wang Zhaoming, et.al, Beijing: Electricity Industry Press. (in Chinese)(1988)
4. J. Dieterich, Constitutive properties of faults with simulated gouge. Mechanical behavior of crustal rocks. A.G.U. Monograph 24: 103-120,(1981).
5. Gao Weiming, et al., The boundary of seismologically risk regions and active tectonics. Earthquake (2). (in Chinese)(1996)
6. Geng Luming, et al., Study on the relationship between the field and the source during the preparation and the occurrence of earthquakes. Earthquake Research in China, 9(4):310-319. (in Chinese) (1994)
7. Li Li, Zhang Guomin and Shi Yaolin, A dynamic model for continental earthquakes, Seismology and Geology, Vol.18 Suppl. Jun.: 9-20, (1996)
8. Qiu Jingnan, et al., Discussion on the stages of seismicity in this century in China. Earthquake (6):41-47. (in Chinese)(1986)
9. J.R. Rice, et al., Dynamic motion of a single degree of freedom system following a rate and state dependent friction law. J.G.R., 91:521-530, (1986)
10. Zhang Guomin, et al., Computer models for the cycle activity of earthquakes in China mainland, Earthquake Research in China, 9(1): 20-30.(in Chinese)(1993)
11. Zhu Yuanqing, et al., Dynamically nonlinear models in the research on seismicity. A.G.S, 34(1):20-31. (in Chinese)(1991)

Milton Keynes UK
Ingram Content Group UK Ltd.
UKHW040106071024
449327UK00019B/858